Aquariculture Biotechnology

The Editors

Dr. S. Felix, Professor of Tamil Nadu Fisheries University at Chennai has been in teaching and research for the past 27 years in the field of Aquaculture Biotechnology. He has developed a number of 'farmer-friendly aquaculture systems' and 'BMPs. He has half a dozen patents registered in innovative aquaculture systems, protocols and new products to his credit. His area of research includes raceway technology, microbial flocculent based aqua farming, designing of cages and lined farming systems, micro algae based biofuel production, fermented biofloc systems and Marine Single Cell Detritus (MSCD) production. He has to his credit over a hundred research papers and reports, seven books and two dozen manuals. He has created important Research farm and Lab infrastructures for his University viz. Fish Biotechnology Centre (Thoothukudi), Maritech Research Centre (Tharavaikullam), Research Farm facility for Aquariculture (Madhavaram) and Raceway Farm Complex (Thoothukudi). His current interest is to contribute sincerely for the development of 'Freshwater Ornamental fish culture sector' a vibrant and the most potential area under aquaculture for the benefit of farming community of the country.

Dr. Arun S. Ninawe, born on 21st June, 1958, is a distinguished scientist holding M.Sc., Ph.D. (Marine Sciences) qualifications. He deals with programs in life sciences and biotechnology under the Department of Biotechnology, Ministry of Science and Technology, Govt. of India and actively involved in coordination and planning activities in identifying research priorities, project formulation, monitoring, evaluation and deals with various policy issues of programme coordination and research funding. He has been dealing this specialized area of fisheries and marine biotechnology, for more than two decades. He is also involved in social sector of the department of biotechnology by supporting programme areas through biotech-based programs for societal development. Presently he is functioning as Scientist "G" and Adviser in the Department of Biotechnology. He has also successfully discharged the responsibility as the Vice-Chancellor, Maharashtra Animal & Fishery Sciences University, Nagpur, during 2007-2010 and supported the academic sector well to give boost to the education, research and extension in Animal Sciences.

As a subject matter specialist he is functioning on various technical committees and professional bodies for promotion of biotechnology in the country.

As a policy facilitator, he has promoted research priorities, identification and formulation of projects in various disciplines of life sciences especially in Marine Biotechnology and various projects have been materialized on bio-prospecting, genomics, aquaculture technology development, health management, fish nutrition and reproduction. He has given emphasis on product and process development in various aquaculture and marine sector and supported Human Resource Development training to benefit fisheries scientist through in service training programme in molecular biology and biotechnology. With his active participation he has been instrumental in funding more than 600 R&D and outreach programmes through training and demonstration activities in the various sectors of animal sciences including aquaculture and marine biotechnology and in various other intra-disciplinary subjects.

Dr. Ninawe has contributed more than 130 research/peer review papers, reports and 8 books to his credit in the discipline of Aquaculture & Marine Sciences field. He has been awarded fellowships of various scientific Societies of India for his valuable contributions in the field of Aquatic and Marine Biotechnology.

Aquariculture Biotechnology

— *Editors* —

Dr. S. Felix, *M.F.Sc. Ph.D.*
Professor and Head
Madhaaram Campus
Tamilnadu Fisheries University,
Madhavaram Milk Colony,
Chennai – 600 0051, Tamil Nadu, India

Dr. Arun S. Ninawe, *Ph.D.*
Adviser/Scientist "G"
Department of Biotechnology,
Room No. 6–8th floor, Block II, CGO Complex,
Lodhi Road, New Delhi – 110 003

2014
Daya Publishing House®
A Division of
Astral International Pvt. Ltd.
New Delhi – 110 002

ISBN 9789351302728

Published by : **Daya Publishing House®**
A Division of
Astral International Pvt. Ltd.
– ISO 9001:2008 Certified Company –
4760-61/23, Ansari Road, Darya Ganj
New Delhi-110 002
Ph. 011-43549197, 23278134
E-mail: info@astralint.com
Website: www.astralint.com

Laser Typesetting : **Classic Computer Services**, Delhi - 110 035

Printed at : **Replika Press Pvt. Ltd.**

PRINTED IN INDIA

Acknowledgement

We acknowledge the overwhelming support and co-operation extended by all the leading Scientists who have contributed chapters for this book:

1. **Dr. Saroj K Swain**

   Principal Scientist, Central Institute of Freshwater Aquaculture, (Indian Council of Agricultural Research), Kausalyaganga, Bhubaneswar – 751 002

2. **P. Nammalwar**

   Institute for Ocean Management, Anna University, Chennai – 600 025

3. **Dr. M.K. Anil**

   Senior Scientist, Central Marine Fisheries Research Institute (CMFRI), Vizhinjam, Kerala – 695 221

4. **Dr. S. Felix**

   Professor, Fisheries Research and Extenstion Centre, TNFU, Madhavaram Milk Colony, Chennai – 600 051

5. **Dr. K. Kumanan**

   Professor and Head, Department of Animal Biotechnology, Madras Veterinary College, Vepery, Chennai – 600 007

6. **Dr. Dhinakar Raj**

   Professor, Department of Animal Biotechnology, Madras Veterinary College, Vepery, Chennai – 600 007

7. Dr. B. Ahilan

Professor, Department of Aquaculture, Fisheries College and Research Institute, Thoothukudi – 628 008

8. Dr. T. Francis

Associate Professor, Department of Fishery Biology, Fisheries College and Research Institute, Thoothukudi – 628 008

9. Dr. Cheryl Antony

Associate Professor, Fisheries Research and Extension Centre, TNFU, Chennai – 600 051

10. Dr. Samuel Masilamoni Ronald

Associate Professor, Bacterial Vaccine Centre, , TANUVAS, Chennai – 600 051

11. Dr. A.S. Ninawe

Adviser/Scientists 'G', Department of Biotechnology, Govt. of India, New Delhi – 110 003

Preface

Skilled manpower development in

'Aquariculture Biotechnology', the need of the hour.......

Despite the tremendous potential, the ornamental aquariculture sector in India has started blossoming only very recently. While this sector needs to go a long way to witness anything closer to **'aquatic rainbow revolution'** in India, things are happening at a faster phase towards the development of this sector is a positive sign! The scientists, fisheries administrators, farmers and the government have realized the potential of this sector at last and genuine attempts are being undertaken to restore this sector which has grotesque marketing potential both in the local and export arena. Further, at a time when SHG concept is fast picking up in our country the women power particularly can be channalised in the sector to reap bumper harvests interms of ornamental fish production.

However, for all these things to happen, a sound technical manpower is the need of the hour, to transform this sector into a economically viable venture in our country. Improved aquaculture systems using bioprocessing methods such raceways, recirculation systems, effective bio-filteration technologies, supplementing live feeds with micro enriched and encapsulated feeds, genetic manipulation to produce innovative colours, diagnostics using molecular and immunological techniques are to name a few advanced biotechnological techniques that can be applied in this sector successfully.

The DBT sponsored Refresher Course on **"Innovative and Viable Biotechnological Techniques for Ornamental Aquaculture System Management" (IBOAS'10)** was organised by our Centre for the benefit of teachers, scientists and scholars to train them on certain important biotechnological techniques of tremendous scope in ornamental aquariculture so as to pave way for disseminating the techniques to the needy sector. The lectures by the leading scientists who have contributed immensely to the filed of 'aquariculture and animal biotechnology' have been arranged for the trainees of the course.

This book is the compilation of these valuable lectures delivered on the occasion and I am highly indebted to the Departmeent of Biotechnology, Government of India who have funded for organising the event in our University.

I am hopeful that the compiled volume of lectures could be of great help to young scientists and scholars to steer their carrier to pursue research in areas of vital significance in the ornamental aquariculture sector.

Dr. S. Felix

Dr. Arun S. Ninawe

Contents

Part IV–Genetically Modified Organisms

Part V–Natural Colour Development in Ornamental Fish

Part VI–Fish Nutrition in Aquariculture

Part VII–Fish Breeding Technology

Part VIII–Advanced Biotechnological Techniques in Aquariculture

Part IX–Fish Health and Diagnostic Biotechnology

Part I

Fisheries Biotechnology: Current Scenario

Chapter 1

Marine Biotechnology: Future for Indian Bio-Economy

*A.S. Ninawe**

Department of Biotechnology, New Delhi

Introduction

Biotechnology is defined as the biological systems which are controlled, manipulated or modified the products to value added products. The impact is previously observed in many industries like food and beverages where it cause many changes in the products and the way they are produced. It is also involved in the development and application of new production system in industries where the production of pharmaceuticals, other chemicals, food and novel materials with application and novel approaches are involved for the processing of waste materials. Many prospective emergence of biotechnology including stem cell therapy and gene silencing technologies are transforming the bio-economy prospects in the scale of trillions. The advanced biotechnology emerges maximum diversity in business sector. The unutilized and unexplored marine resource is the important biological source which is beneficial for industrial sectors. In Europe, bio-economy was established which utilize a biological resource and it estimates around 22 million employee yields a market size of over €1.5 trillion.

Marine Biotechnology is the creation of products and processes from marine organisms through the channel of biotechnology, molecular and cell biology, genetic engineering, rDNA technology and bioinformatics. It is a knowledge generation and

* *Corresponding author.* E-mail: ninawe@gmail.com

conversion technique which utilizes biological compounds and makes use of them. It is a very fascinating and economically expanding field provides greater genetic diversity with high possibilities of natural products. More than 80 percent of the earth's living organisms are found only in marine ecosystem with little known for their biochemical characteristics and provide opportunity to discover the marine micro organism for producing biologically active substances. Understanding of biodiversity of marine resources of symbiotic nature of marine organisms increased the interest of marine biotechnologist to addresses the worldwide challenges includes food security, fuel security, population health, green growth and sustainable industries and also ecosystem services for planet as food, regulators of global temperatures, pollution filters, and nutrient and mineral cycling. Therefore, marine biotechnology has a wide scope for various applications.

The marine bio-economy refers to the sustainable production process and/or utilization of marine bio-resources for a range of food products, healthcare leads, energy, and pharmaceuticals. The marine bio-economy include aquaculture and processing industries and other industrial sectors making use of biological resources (*e.g.*, seaweeds) and fisheries. The overall attractiveness of marine technology to businesses outside the marine sector is the highlighted successes of the potential and their strength. From the last two decades various bioactive natural compounds which are secondary metabolites, seeking a lot more attention from chemists and pharmacologists have been used in food, fragrances, pigment, insecticides and medicines.

Marine Biotechnology unravels with marine resources to solve problems discovering new compounds having therapeutic and industrial applications. It also gives equal importance in the need for new tools to monitor ecosystem health and find solutions for difficult environmental problems. Recently a lot of efforts could be taken for development of new drugs of life threatening diseases such as cancer and AIDS from marine resources and use of such significant life forms act as an important strategy in marine biotechnology. The role of these products in drug discovery has been greatly enhanced in the last few years through the improved biological screening method and has opened up for diverse and novel chemical structures with potent biological activities. The cutting-edge techniques could be useful for accessing marine microbes which are yet to be cultured and thereby increasing the microbial resources that target for drugs.

Growth of Biotechnology and Marine Biotechnology Sector

The Indian biotechnology sector is one of the fastest growing knowledge-based sectors in India and is expected to play a key role in shaping India's rapidly developing economy. With numerous comparative advantages in terms of research and development (R&D) facilities, knowledge, skills and cost effectiveness, the biotechnology industry in India has immense potential to emerge as a global key player. India has been ranked among the top 12 biotech destinations worldwide and third largest in the Asia-Pacific region, addressing the key segments in the Indian biotechnology industry *viz.* Bio-pharmaceuticals, Bio-services, Bio-agriculture, Bio-industrial and Bio-informatics. The Indian biotech industry registered 18.5 per cent growth in FY12; total industry size stood at US$ 4.3 billion at the end of the financial year. The industry is expected to grow to US$ 11.6 billion by 2017, driven by a range of factors including growing demand, intensive R&D activities and strong government initiatives. The bio-pharmaceutical sector accounted for the largest chunk of the biotech industry, with a share of 62 per cent in total revenue in FY12. In the same year, bio-services and the bio-agri segments followed the bio-pharmaceutical segment with shares of 18.3 per cent and 14.9 per cent respectively. Growth was fastest in the bio-agri segment (23 per cent in FY12).

Marine Biotechnology and the Food Sector

To enhance the visualization of India as a knowledge-based economy in the sectors of marine-foods and products over the innovation driven culture demands supports from the state of encouragement and expands research activity. The focus of marine biotechnology was diversified with different funding agencies such as ICAR, DRDO (*e.g.* bio-fouling) etc., whereas the Department of Biotechnology is promoting this sector through an exclusive Task Force on Aquaculture & Marine Biotechnology. Innovations in mariculture technologies coupled with greener ocean efforts would encompasses a sustainable food production system which could meet the global requirement of protein-rich foods to eliminate malnutrition. As there is a clear requirement to tailor animals and plants to meet requirements of the protein-rich farmed foods and efforts to ensure ocean stock even through industrial / over fishing can keeps marine foods as sustainable.

Non-food Applications

The marine products that are been harvested in high value bio-actives for the application in human health and animals, including pharmaceuticals, nutraceuticals and functional foods are rapidly progressing. Having known these bioactive compounds from marine organisms, an additional scientific challenge is to develop the capability to tailor marine species to produce higher and purer amounts of specific bioactives. System biology, bioinformatics and high throughput screening formats are developed in demanding expertise as issues of reliability and extraction efficiencies, stability and even attracting the bioactivity of specific molecules. While such a huge scope considering the potentials of growth of biotech industry in India, development of Marine Biotechnology is at its nascent stage. One of the major constraints is lack of trained manpower and adequate infrastructure facility.

Recent Advances and Global Bio-business

The impact of biotechnology processes is generated over the next generation products and the requirement of the scaled-up processes to unlock the anticipated valuable materials of these sectors. The marine is diverged not just in biodiversity term as it exists in it but the term exists in respect of the scientific skills and expertise required to understand, harness and development of opportunity in the enterprise. There are hidden opportunities at all levels of ocean environment for "discovery science". The terrestrial agri sector was fully over shadowed by the potential impact of the marine biotechnology sector despite the fact only a small part of the marine biosphere has been sampled. Recent expedition by Grag Venter's team in the Sargasso sea revealed novel genera and lineages of marine microbes (Venter *et al.*, 2004). Thus metagenomic exploration of marine environment would bring out new industrial leads which could not be access with conventional methods.

Marine biotechnology explores the potentials of marine organisms at cellular or molecular level, to provide inside in to existing problem with the use of technology to advance the understanding and accessibility of marine biological resources in improving the life sustainability by restoring and protecting aquatic ecosystem, enhance the food supply through aquaculture, seafood safety and quality enhancement, development and processing of new types and resources of industrial materials, expand knowledge of biological and geochemical processes in the world ocean. It has high value global market for Marine Biotechnology products, processes and estimated at € 2.8 billion (2010) with a cumulative annual growth rate of 4-5 per cent. Less conservative estimates predict an annual growth in the sector of up to 10-12 per cent in the coming years, revealing the huge potential and high expectations for further development of the marine biotechnology at a global scale. Therefore marine bio-resources offer a lot of scope for the health and well being of aquaculture production in the country. Recent focus on marine biotechnology world-wide is through bio-discovery of marine microbes, invertebrates, micro-algae and macro-algae which are likely to play a major role in the commercial production of bio-active molecules and pharmaceuticals. Among the marine organisms, seaweeds and marine sponges were the most explored potential candidates for drug discovery and bio-actives (Figure 1.1).

Figure 1.1: Some of Marine Sponges and Seaweeds Collected from Indian Coast and Tested for Bioactivity and Aquaculture Drugs.

Marine Bio-fuels

There has been a stimulated interest over energy crop growing as a result of steady increase in world energy prices and its association over the increase of bio-fuel demand. One promising source of biofuels has been identified as marine algae cultured in large open ponds. The algae would be harvested and processed into a carbon neutral fuel source. Along with camelina and jatropha, marine micro algal biomass is a front runner in the search for a renewable feedstock for production of commercial quantities of biofuel. While marine algae produce lipid oils, fat molecules that store energy needed for fuel production, the process occurs under nutritional depletion which culminates their growth. When well-nourished, the algae do grow well, but produce carbohydrates instead of the desired lipids for conversion to fuel. A current breakthrough by Trentacoste *et al.* (2013) evident that metabolic engineering to target a specific enzyme secreted in diatoms can trigger the production of lipids (biofuels) without hurting growth.

Research Caliber Development

The marine biotechnological research established for developing new drugs, human- well being, finding novel enzymes, bioactive compounds that could be synthesized and manufactured as pharmaceutical products or biomedical research tools, biopolymers and metabolites for wide range of industrial applications, biosecurity of aquaculture and fisheries, etc. The versatility of marine products has huge applications in various industries to develop new class of pharmaceuticals, vaccines, diagnostics, analytical reagents, industrial products and processes. There has been a rapid increase in the inventory of marine natural products and genes of commercial interest derived from bio-prospecting. While understanding the focus of genomics, currently, about 1000 prokaryotic genomes have been sequenced and annotated. More than half of these genomes are of medical or industrial relevance and no phylogenetically systematic genome sequencing is being carried out resulting in to the discovery of many novel proteins and demonstrating the existence of a huge reservoir of undiscovered proteins. The R&D leads on utilization of marine genetic resources contributed over 18,000 natural products, 4,900 patents associated with genes of marine organisms with greater effort on sustainable development.

Research Centers like PIBC (Pacific Institute of Bioorganic Chemistry and NOAA Research investments working on 8000 strains of bacteria with developments as new anti-cancer drugs from marine invertebrates and genetically engineered microbes for use in oil spill cleanup, synthetic antifreeze, water-resistant adhesives, and super absorbent materials from various marine organisms. Researchers have isolated and identified marine bacteria that are capable of degrading PAHs and sources for industrial production of many enzymes. Harbor Branch Oceanographic Institution's Biomedical Marine Research Division found out valuable products from marine specimens including microorganisms associated with deep water sponges and other deep-water invertebrates and its application in cancer treatment. Sea Grant researchers attempting on product development from marine organisms and are discovering many life-saving drugs. A compound derived from mangrove tunicates and corals shown to have a potent anti-tumor treatment and anti-cancer activity. NOAA Research

Office of Ocean Exploration supporting several expeditions to search deep-water habitats for marine organisms that may contain bioactive compounds that focusing emphasis on synthesizing and manufacturing as pharmaceutical products or biomedical research tools.

From the foregoing it is seen that there are immense potential for biotechnology in marine sciences. The greatest need of today is for exploitation of biologically active compounds and the hither to untapped feed resource. India has a vast coastline of about 8000 km and its Exclusive Economic Zone (EEZ) extends upto 370 km from the coast. Appropriation of marine bio-resource distributed within a vast, complex, dynamic and shared ecosystem, its protection and preservation for future generations is a challenge for marine biotechnology sector. These challenges need alignment across the national, regional and international levels by strengthening collaborative research between academic research institutes and industry. The fair and equitable access to marine genetic resources with innovative approach can boost up the growth with creation of stronger identity and communication through awareness building in marine biotechnology research.

India Needs Strategic Approach

The biotechnological strategies needed in marine aquaculture activities are the farm engineering and harvesting technologies which have to be designed and perfected for production of maximum yield. Chromosomal manipulation techniques may be utilized and perfected to improve fish production in marine aquaculture. Fish nutrition is another aspect which has to be considered from the point of view of marine aquaculture. The feed for the organisms to be reared has to be carefully compounded with due regard for increased growth rate and breeding so that their culture is profitable. The pathology and diseases of organisms under marine aquaculture have to be studied and their control using genetic engineering techniques as well as immunological methods found out. The post-harvest technologies to increase the quality and value of the final feed products have to be improved. Another very important area which needs our attention is the transportation of seeds of the organisms from the hatchery to the farm or field for marine aquaculture. Today the transportation of seeds present a risk of mortality which is very high. Suitable and specified technologies have to be designed and perfected for each organism so that the mortality rates can be kept minimum and that the seeds survive for longer duration. Further, biotechnology technique can be applied for the creation of new and improved strains of cultivable species. These would grow faster and larger, contain more edible fraction, can grow within an expanded range of salinity, are disease resistant, amenable for mass culture and possess many other advantages. But all these can materialize only if more basic information is available on the physiology, biochemistry and genetics of cultured species and on refining the existing biotechnological techniques.

Promotion of Marine Biotechnology

In India, the technological research in the area of aquaculture and marine science is quite recent. It holds great promise in the areas of marine bio-medicals, marine toxins, marine industrial chemicals, mariculture, marine microbiology, marine bio-fouling, marine-pharmacology, marine environmental management, marine pollution and others. India to compete with the developing world in cutting research areas we need to explore the marine life that permits direct utilization of new and unique products both by genetic engineering to ensure continuous production in the laboratories and bio-incubators, eliminating fluctuations caused by weather and climatic conditions.

Department of Biotechnology (DBT) is laying emphasis on marine biotechnology through its Aquaculture & Marine Biotechnology programme covering both research and applied aspects covering support through Competitive Grant Support with the Task Force Committee/ Brain Storming and Workshops. During the last two decades, the department has been concentrating on developing technology for seed production and grow-out pond culture for economically important finfish and shellfish species, seaweeds and development of commercially and economically important species. With its support it has been possible to produce fresh water prawn up to 1.5 to 2.0 ton/ha. Farmers have adopted this technology for flattening and realizing good returns. In aquaculture shrimp culture through semi intensive farming it has been demonstrated to produce 10 t/ha per annum in two crops and demonstrated viable satellite farming system is being developed with good farm inputs of seed, feed and the best management practices. It has also been possible to produce 18-20 t/ha in fish production per annum through poly-culture with introduction of high yielding hybrids, quality feed and breed and better health management. This has been a paradigm shift in fish and shrimp production in the country.

Research on induced breeding and maturation in commercially important fish and shrimp species is being carried out. Research has also been focusing attention on various aspects such as diagnostics, vaccines, recombinant products, anti-microbial peptides, immunostimulants, high-energy and high-protein aquaculture feed, fish spawning agents, human therapeutics (omega-3 and 6), water quality management, value addition and product and process development. Inventorisation and digitalization of marine and coastal bio-resources pertaining to marine mollusk are being carried out apart from introduction of Public-Private Partnerships were pursued for leads in water re-circulatory system in marine biotechnology especially in aquaculture health management by controlling metabolite load in the culture operations etc. During the last decade the research has been focused on aquatic health management with emphasis on development of technology for screening healthy population of shrimp/fish brooders based on molecular markers. Application of various technologies especially biotechnological tools has made an impact in reducing disease risk and contributing to the future enhancement of aquaculture production. Efforts were directed towards the development of a number of conventional and molecular diagnostics and therapeutics and by enterring in to joint collaborative projects on development of vaccines against viral and bacterial pathogens encountered in Indian aquaculture systems. Several sensitive and specific diagnostic assays have been developed to detect viral pathogens such as white spot syndrome virus (WSSV), fish nodavirus (VNN) and other pathogenic viruses. The successful diagnostics developed for diagnosing WSSV and white tail disease (WTD) associated

Macrobrachium rosenbergii nodavirus (*Mr*NV) and extra small virus (XSV) are available as health management measures in aquaculture.

Vaccine is very important to prevent diseases in aquaculture systems and with its sponsored projects DBT has developed vaccines against different viral and bacterial pathogens of fish and shrimp. Recombinant and DNA vaccines have been found promising for *Aeromonashydrophila, Vibrio anguillarum, Edwardsiella tarda,* fish nodavirus, WSSV and *Mr*NV. Recombinant vaccines using different genes such as OmpTS, Aha1 and OmpW were developed and their efficacy against *A. hydrophila* have been found to be effective. Inactivated whole WSSV vaccine, recombinant subunit vaccine and DNA vaccine using different viral genes have been proved to be useful in small scale against WSSV in shrimp. Protection of *Fenneropenaeus indicus* from WSSV has been demonstrated using formalin-inactivated WSSV. Different types of immunostimulants of bacterial or plant origin have been developed for cultivable organisms for their protection from viral and bacterial pathogens. Laboratory studies indicate that administration of a commercial Aquastim MBL by immersion or through feed stimulates the immune system of shrimp in hatchery, showing significant improvement in survival.

Application of RNAi to control WSSV in shrimp has shown good leads to produce WSSV-free shrimp brooders and seeds. Tissue culture and the development of cell lines from fish are priorities for pathogen detection, toxicological studies, carcinogenesis, cellular physiology, and genetic regulation and expression studies. Programme supported on fish cell lines from marine and freshwater fish at different research institutes in network mode and more than 50 fish cell lines from economically important species developed. These are being established at the National Repository of Fish Cell Lines (NBFGR, Lucknow) for future use and application.

In aquatic health management, technology is being developed for screening healthy population of brooders based on molecular markers, maturation in captivity, completion of maturation in indoor facility wherever feasible, breeding and spawning in controlled systems.

Novel concept on raceway technology using aerobic microbial flocculent technology is being propagated in designing, construction and operation of raceways to raise crops of various kinds including shrimp, scampi and freshwater ornamental fishes for future and to ensure the bio-security of used water for practicing sustainable aquaculture.

The department of biotechnology is addressing various aquatic health issues through establishment of diagnostic laboratories to screen the shrimp brooders and seeds for viral pathogens using molecular tools such as PCR, RT-PCR and immunodiagnostics through the projects supported at various aquaculture institutes in the country. Support is also provided to the programmes like production of SPF brooders and seeds using molecular tools including RNAi technology and developing brood stock banks to supply of high quality brooders to hatchery operators. Fish cell lines and shrimp primary cell culture, the outcome of DBT-funded projects, are being used for propagation of viruses for production whole virus vaccines. There is a need of National Referral Facility to take care of aquatic animal health by providing diagnostic facility and control and preventive measures for all bacterial and viral diseases encountered in Indian aquaculture system.

The department has taken a major initiative in strengthening the infrastructure and facility to bring about academic excellence in Aquaculture & Marine Biotechnology both in basic and applied aspects. During the 10th plan final year, the Department has provided support to two institutions by sanctioning a dedicated programme support on Aquaculture & Marine Biotechnology at College of Fisheries, Mangalore, and Cochin University of Science & Technology by giving major emphasis on the creation of a Centre of Excellence in the area of Marine Biotechnology addressing important areas like RNAi technology, nano-technology, extremophiles, cell-line development and their application, user friendly diagnostics and vaccines for new emerging diseases, immuostimulants, brood-stock development, production of disease free seeds, low cost feed development, and post harvest technology, natural product development, bio-fuels from marine algae, novel microbial enzymes for industrial applications, novel drugs from marine organisms for medicinal value etc. Further, aquaculture and marine biotechnology sector has been greatly benefitted with the funding from industry promotion schemes of DBT, SBIRI and BIPP through products and process development having high value export potential through active industry participation.

Key Areas for Funding Support

The department has identified integrating advanced biological chemical and engineering platforms with Marine biology to provide means to address pressing issues in marine biotechnology including health management and productivity of cultured organisms. Bioactive compounds and pharmaceuticals as the source of novel metabolites particularly marine actinomycetes, have been extensively studied for many bioactive compounds. These substances have application in the treatment of a wide variety of diseases. Some of the most promising opportunities on enzymes produced by marine bacteria display unusual properties such as salt-resistance - a characteristic that is often advantageous in industrial processes; extra-cellular proteases of particular importance for their use as detergents and in industrial cleaning applications. These are being explored for research promotion. Marine organisms harnessed to produce new biomaterials such as novel biodegradable polymers, biomedical implants and devices. The natural substances in the form the organic matrices of molluscan shells and anti-biofouling agents have immense application potential. Development of biosensors, bio-indicators and diagnostic devices can be applied in a broad range of situations in medicine, aquaculture and environmental monitoring and the developing for biomedical applications and for insights into the basis of disease mechanisms and pathogenesis in humans. Studies of the developmental, cellular and molecular aspects of marine organisms are expected to provide some answers. Further, biodiversity and conservation of genetic resources are essential to identify and characterize and understand important species especially unusual or endangered ones. There are several believable constraints of not getting adequate support for biotechnology research due to various policy issues by meeting these challenges to identify sources of ecological stress, protect and restore coastal resources and promote research in marine biotechnology.

References

Anon. 2013. Annual Report, Department of Biotechnology, Govt. of India, pp: 200.

Das, P. 1991. Genetics in enhancing aquaculture productivity. In Aquaculture Productivity (Eds. Sinha, V.R.P., H.C. Srivastava), Oxford and IBH Publishing Co., New Delhi. pp: 623-627.

Iddya Karunasagar, Indrani Karunasagar (Editor), Alan Reilly (Editor). 1999. Aquaculture and Biotechnology: Current Development, publisher: Science Pub. Inc., pp:186.

James, P.S.B.R. and Ponniah, A.G. 1991. Concepts in marine biotechnology and their application for enhancing aquaculture productivity. In Aquaculture Productivity (Eds. Sinha, V.R.P and Srivastava, H.C.), Oxford and IBH Publishing Co., New Delhi. pp: 559-605.

Lakra, W. S., Abidi, S. A. H., Mukherjee, S. C. and Ayyappan, S. 2009. Fisheries Biotechnology, Narendra Publishing House, pp: 240.

Mary-Frances Thompson, Rachakonda Sarojini, Rachakonda Nagabhushanam. 1991. Bioactive compounds from marine organisms. Oxford and IBH Publishing Co.Pvt.Ltd. New Delhi. pp: 410.

Ninawe, A.S. 2006. Biotech Promotion Policy and Technology Transfer Issues: Evidences from Aquaculture and Marine Biotechnology, Asian Biotechnology and Development Review Vol. 8 No.3, pp: 17-33, RIS. All rights reserved.

Ninawe, A.S. 2006. India's endeavors in biotechnology : A policy overview, Tailoring Biotechnologies Vol. 2, Issue 3, Winter 2006/07, pp: 107-130.

Ponniah, A.G., Das, P. and Verma, S.R. 1988. Fish genetics and Biodiversity conservation. Nature Conservators, Muzaffarnagar, India. pp: 474.

Trentacoste *et al.* 2013. Metabolic engineering of lipid catabolism increases microalgal lipid accumulation without compromising growth. PNAS (In press).

Venter *et al.* 2004. Environmental Genome Shotgun Sequencing of the Sargasso Sea. Science 304: 66-74.

Chapter 2

Present Status of Marine Biotechnological Research and Development in India

P. Nammalwar

Institute for Ocean Management,
Anna University, Chennai – 600 025, T.N., India

Introduction

The concept of biotechnological research in the fields of aquaculture and marine science is quite recent. These fields hold vast potentials for biotechnological research in the areas of marine biomedicals, marine toxins, marine industrial chemicals, marine aquaculture (or mariculture) marine microbiology, marine- biofouling, marine pharmacology, marine environmental management, marine pollution and others. Marine biotechnology holds the greatest promise for developing countries where fish and shellfish are a major source of feed, as well as an industrial commodity. Further more, developing countries possessing abundant marine and estuarine natural resources are ideally suited for exploration of marine biotechnology, since access to unusual and/or novel marine life permits direct utilization of new and/or unique products by genetic engineering and ensures continuous production in the laboratory, eliminating fluctuations caused by weather and climate.

Marine Biomedicals

It is now known that several marine organisms contain many bioactive chemical compounds with various pharamacological properties. Many marine organisms have provided useful drugs. These liver oil from fist provides excellent sources of vitamins

X and D; insulin has been extracted from whales and tuna fish; and the red alga Digenia simplex has been shown to contain an anthelminthic.

An outstanding example of the potential biotechnological application offers is that of marine pharmaceuticalsand drugs. Cardiotonic polypeptides have been extracted from sea anemones, adrenergic compound from sponge, potential anti tumor compound from Caribbean gorgonians and soft corals. Antiviral and anti tumor despeptides have been obtained from a Caribbean tunicate of the family Didemnidae, which inhibits the growth of DNA and RNA viruses as well as L 1210 marine leukemic cells. There are indications that tunicates may be an abundant source of bioactive compounds. Cardio vascular active substances have been isolated from sponges, N-methylated histamines and histamines from Verongia fistularis, asystolic nucleoside from *Dasychalnia cyathnia* and the nucleosids spongosine from *Cryptotethya crypta*. These provide a few examples.

The fact remains that it is uneconomical to extract and purify material from organisms that have to be captured in large quantities from various remote areas. This necessitates the need for culture of these organisms. Further there is a lack of knowledge concerning the basic chemistry of many of the marine natural products which has limited sources for the development of useful drugs. Genetic engineering can change this situation dramatically by revealing the vast and diverse genetic composition of marine life for pharmacological application.

Marine Toxins

Many powerful chemical toxins have been isolated from various toxic organisms. They have specific functional groups in the molecule and show strong toxic physiological activity. Many toxins have potential applications as a drug or pharmaceutical agent. Even when the direct use as drug is not feasible, because of potent or harmful side effects, these toxins still serve as models for synthesis or providing suitable derivatives which improves their suitability as drugs.

Tetrodotoxin found in pufferfish (Tetrodontidae) can paralyse peripheral nerve and is valuable, because it inhibits the sodium permeability of nerve membranes and is useful for elucidating the excitation mechanism. Both saxitoxin (obtained from clams *Saxidomas gigantex, Mytilus* sp. and *dinoflagellates Conyaulaux* sp.) and tetrodoloxin have been used to study the structure of sodium channel of membranes.

Nereistoxin was isolated from the marine annelid *Lumbriconereis heteropoda* which is commonly used as bait the toxin is an insecticide and deadly to flies. Once its structure was established, a new insecticide PADAN was developed from nereistoxin. Cartap hydrochloridxe is one of its synthesized derivatives which is active against the rice stem borer and other derivatives which is active against the rice stem borer and other insect pests. It is not toxic to warm blooded animals and the resistant strains of the insects do not develop readily.

An extensive literature providing information on marine organisms causing feed poisoning or having capacity to produce toxic or poisonous reactions by sting or bite is now available. The vast array of marine animals and plants are covered. These include algae, coelenterates, echinoderms, corals, sea anemones, molluses, flagellates, nemertines, annelids and other invertebrates and fish. The research on marine toxins or bioactive substances in marine organisms has increased in the recent years and a number of monographs and reviews are available.

Palytoxin an extremely poisonous water soluble substance has been isolated from zaonthid coral of genus *Palythoa* and it structure elucidated. This toxin with its stereochemical structure represents an entirely new class of natural products. It is biogenesis is not clear. It's similar in potency to rice, a polypeptide toxin found in castor beans. It exerts it lethal effect when administered intravenously in low concentrations to laboratory animals. It influences calcium and potassium and iron transport in nerves and the heart. Animals undergo paralysis and heart failure.

An interesting feature of palytoxin is that it is synthesized by a marine Vibrio species growing symbiotically with the coelenterate Palythoa and apparently related to *Vibrio cholerae*. This Vibrio is only a mild pathogenic for human causing only influenza like illness and is rapidly attenuated in laboratory culture, losing its ability to make toxin.

Halitoxin, a toxin mixture from several marine sponges of the genus *Haliclona*, has been isolated and proved to be of a complex mixture of high molecular weight. It is toxic for fish and mice and also inhibit the growth of Ehrlich ascites tumors. It may prove to be an anti-tumor agent. Similarly, lophotoxin a new neuromuscular toxin isolated from several gorgonians of the genus *Lophogorgia* has been isolated. It inhibits nerve-stimulated concentration without affecting contraction evoked by direct electrical stimulation of the muscle. Another interesting compound, bryostatin which has antitumor properties, has been isolated from marine animals of the phylum Ectoprocta specifically from *Bugula neritina*. The extremely low dose for antineoplastic activity suggests that bryostatin may have other potentially useful pharmacological or microbiological activities, sufficient to merit its use as a biochemical probe.

Hybridoma technology offers a valuable tool for characterizing the marine toxins. The application of this technology is marine pharmacology are practically unlimited, including study of the structure and functions of the toxins, as well as for the production of antioxins for treatment. By this technology monoclonal antibodies to the enterotoxin of *Vibrio cholerae* has been produced. Vibrio cholera is a brackishwater and estuarine bacterium causing cholera and apparently related to the *Vibrio* sp. producing palytoxin. Monoclonal antibodies against cholera toxin were produced to obtain highly specific antiserusm to cholera toxin.

The hybridoma approach to characterise these compounds should permit wider screening for the distribution of the toxins among various sponges and lead to subsequent isolation and testing.

The application of marine toxins are limited and it is mainly used at present in the area of understanding functions that the toxins have.

At present, strategies are needed for collecting, culturing and screening of the marine organisms from which bioactive agents can be isolated and characterized. The need to study the fundamental chemistry, to determine the structures of new compounds cannot be over emphasized. The immediate successes are not likely to occur with discoveries of novel antineoplastic agents antibiotics and anti-inflammatory agents produced by marine bacteria, plants and animals.

Marine Industrial Chemicals

Marine polysaccharides, carotenoids, unusual sugars, enzymes and algal lipid from the sea may come as marketable products. Seaweeds form as source of raw materials for numerous industries. Carrageenan, obtained from red algae is widely used as extender in feeds and related products, such as evaporated milk, toothpaste, etc. Agarose, another polysaccaharide is widely used in electrophoresis and chromatographic analysis in the laboratory. Most of us today consume directly or indirectly or come in contact with some form of algae daily. Many applications of genetic modification in seaweeds to the commercial utilization and culture have been developed and reviewed. In general these technologies and improved techniques to seaweed culture is recent and somewhat limited. The most common and simple technique is strain selection, such as screening of plants describe traits such as fast growth. Protoplast fusion - somatic hybridization techniques are now applied to agar producing seaweeds. The salt tolerant microbial systems provide chemicals such as polysaccharides, enzymes and kinds which have great metabolites because they are frequently abundant. These unusual compounds may be pathway intermediates and also a potential source of new chemicals. Various terpenoids, fatty acids and saponins are found in many of the sponges, gorgonians and corals. Sea-hares contain interesting metabolites often showing to be from algae. The extracts of the sea-hare *Aplysia dactylometa* show cytotoxicity and antitumour activity; and it contains a variety of halogenated metabolites.

Researches on the cloning and expression of sea urchin histone genes have been conducted and the molecular biology of the sea urchin embryo studied recently. The developments in molecular genetics in propelling industrial research into newer areas of activity have begun. A directed search for biologically active natural products in the marine environment particularly from marine organisms can open an entirely new source of industrial chemicals. The need at present is new noval screening strategies for such products.

Mariculture

Most of our traditional fisheries are being harvested from the sea at or near maximum sustainable yields. Fisheries provide food to the people. The food provided by agriculture and fisheries is not sufficient to our country with its huge population and there exists vast deficit. Marine aquaculture provides a good potential to reduce this deficit. Exploiting of the microbial sources of protein at larval stages, during larval metamorphosis and their growth can produce huge benefits and profit. The application of genetic engineering to improve food yield is clearly a very promising area to focus our attention.

Applications of biotechnology for enhancing aquaculture productivity needs work along two lines. One is on basic aspects of physiology and genetics of the species of interest. The studies must be oriented towards elucidation of the mechanisms of control of traits like osmoregulation at the biochemical, physiological and genetic level so that it might be possible to manipulate them using biotechnological techniques. The other line approach has to be on the selection and modification of biotechnological techniques for manipulation of the selected trait the traditional genetics. Technique of selective breeding, inbreeding, interstrain crossing, interspecies hybridization can be selectively combined with more recent techniques like polyploidisation, gynogenesis and monosex culture to increases aquaculture productivity. There is also scope for novel improvements through exploitation of the high fecundity and the flexibility inherent in external fertilization of aquatic organisms.

The applications of biotechnology can lead towards enhancing the growth rate, production of specific bioactive compounds or towards more exotic objectives of producing genetically altered strains of shrimp and molluses cultivable in very low saline or freshwater, reduction in the non- edible portion and producing a strain which can circumvent the marine environment for their breeding cycle.

Marine Microorganisms as a Source of Bio-Diesel

Marine micro algae can provide different types renewable bio-fuels. Microalgae mass cultures have been considered for almost 50 years as potential biofuel sources, with the first conceptual engineering analysis presented in the late 1950s. Initially proposals were for the production of biomass suitable for methane fermentation in combination with waste water treatment (anaerobic digestion). More recently emphasis has been on photobiologically produced biohydrogen. The idea of using microalgae as a source of fuel is not new, but is now being taken seriously because of the escalating price of petroleum and more significantly, the emerging concern about global warming that is associated with burning of fossil fuel. Biodiesel is now essentially produced from plant and animal oils but not from microalgae. This is likely to change as many companies are now attempting to commercialize microalgal biodiesel.

Marine Microbiology

Genetic Engineering has found its widest application in terrestrial microbes. Extension of these techniques to marine microbes should be relatively easy than in other organisms. Photosynthetic bacterial (PSB) found in mangroves and estuaries can be used for biotreatment of waste water due to their ability to remove carcinogenic agents and produce anti-viral substance. Genetic engineering 'techniques can be used to develop strains of PSB to degrade compounds that are not easily broken down. The harvested PSB can be incorporated in fish feed since even at low percentage it enhance fish yield. Also genetic engineering can be used to enhance fish yield. Also genetic engineering can be used to enhance production of anti-viral or anti-biotics from marine microbes.

Marine Biofouling

Marine biofouling is highly destructive to vessels and under water and floating structures used for marine aquaculture. The ability of bacteria to find, attach, adhere and elaborate specific primary films are the crucial stages in biofouling. If these factors involved are understood, it is possible to manipulate them by employing biotechnological techniques. Two approaches are being tested to elucidate the molecular basis of fouling. One is to identify the genes involved in each of this process using recombinant DNA technology. The other is the use of transposon mutagenesis, when a transposon mutant deficient in the expression of adhesion gene is discovered it could be easy for further elucidation of the factors involved in microbial adhesion and then it might be possible to manipulate these factors at a genetic and biochemical level. This has also implication in aquaculture by enhancing spat settlement of cultivable molluscs.

Marine Pharmacology

A dramatic example of biotechnology applications is that of marine pharmaceuticals. Extracts from a tunicate belonging to the family Didemnidae, inhibit growth of DNA and RNA virus as well as leukemic cells. Marine toxins besides being pharmacological chemicals, also serve as models for development of new synthetic chemicals. The strategy to be undertaken in marine pharmacology would be initial screening of marine organism for bioactive agents. The most useful bioactive compound can be tested and characterized. Then a two prolonged strategy could be undertaken. One for evolving techniques for mass culture of the organism and the other for using recombinant DNA technique to identify and clone genes responsible for synthesizing the bioactive compounds for increasing its production.

Marine Pollution Control and Management

Synthetic compounds are relatively resistant to biodegradation compared to natural products. This is more because natural organisms cannot produce enzymes necessary for transformation of the original compounds so the intermediates are produced which can enter into common metabolic pathways and metabolized completely. This creates special problems for waste management and environmental protection.

Groups of microorganism useful in treating specific types of man made compounds have been compiled and are known. Selective use of microorganisms, including actinomycets, fungi, bacteria, phototrophic microorganisms, anaerobic bacteria and oligotrophic bactera are known in certain application such as waste water treatment for biological removal of nitrogen via sequential nitrification and denitrification. The treatment of selected industrial wastes in reactors using controlled mixed cultures is in use in Japan. Various methods of genetic engineering will certainly prove to be of use for this. The engineering of microorganisms to be added to wastes that are to be discharged into the marine environment has to be found out. Pollutants entering the marine environment can interfere with the integrity of the ecosystems. The pollutants include synthetic organic compounds, chlorinated chemicals, dredged spoils, litter, artificial radionuclides, trace metals and fossil fuel compounds. Toxaphene, a group of about 200 compounds produced by chlorination of wood waste products and comphors under ultraviolet light contains carcinogenic and mutagenic chemicals. These may be more persistent in the environment than DDT and its degradation products.

The modifications of genetic information resident in microorganisms can be used for pollution control. The enzyme concentration may be amplified either by selection of constitutive mutants, increase in the number of copies of the gene for the enzyme or both, so that enzymatic degradation can be made. Rearrangement of regulatory mechanisms controlling the expression of specific genes in response to specific stimuli may also prove to be useful. Another useful modification is the introduction of new enzymatic functions into organisms not having them. The alteration of the characteristics, such as substrate specificity, kinetic constant and pH optimum, of specific enzymes can also be made use of. These modifications can be achieved by undertaking invitro modifications via transponson mutagenesis or other transposon mediated gene manipulation, genetic exchange via, transduction, transformation, or conjugation, protoplast fusion, specific site mutagenesis, and specialized selection procedures to enrich for mutants. The engineering of micro-organisms capable of flourishing in the marine optimized proliferation and maintenance of selected populations has to be considered.

The need for algicides and antifouling agents is so great that discoveries of such compounds with these activities will produce marketable success.

Suggestions

From the foregoing it is seen that there are immense potential for biotechnology in marine sciences. The greatest need of today is for exploitation of biologically active compounds and the hither to untapped feed resource. India has a vast coastline of about 8000 km and its Exclusive Economic Zone (EEZ) extends upto 370km from the coast. The pioneering researches conducted by the Central Marine Fisheries Research Institute has shown that there are numerous species of organism belonging to sponges, gorgonids, corals, echinoderms, sea-anemones, coelenterates, alcyonarians, sea-urchins, molluscs, pearl and edible oysters, clams, seaweeds and algae, sea greases, etc. in our coastal waters, Screening of these for their bioactivity and isolation and characterization of the chemical compounds responsible for these activities is the first step needed. Already some researches have been conducted by Central Marine Fisheries Research Institute (CMFRI), Central Institute of Fisheries Technology (CIFT), National Institute of Oceanography (NIO) and Central Drug Research Institute (CDRI), but it is only peripheral.

The other biotechnological strategies needed in marine aquaculture activities are the farm engineering and harvesting technology which have to be designed and perfected for production of maximum yield. Cromosomal manipulation techniques may be utilized and perfected to improve fish production in marine aquaculture. Fish nutrition is another aspect which has to be considered from the point of view of marine aquaculture. The feed for the organisms to be reared has to be carefully compounded with due regard for increased growth rate and breeding so that their culture is profitable. The pathology and diseases of organisms under marine aquaculture have to be studied and their control using genetic engineering techniques as well as immunological methods found out. The post-harvest technologies to increase the quality and value of the final feed products have to be improved. Another very important area which needs our attention is the transportation of seeds of the organisms from the hatchery to the farm or field for marine aquaculture. Today the transportation of seeds present a risk of mortality which is very high. Suitable and specified technologies have to be designed and perfected for each organism so that the mortality rates can be made minimum and that the seeds survive for longer periods. Further, biotechnology technique can be applied for the creation of new and improved strains of culturable species. These would grow faster and larger, contain more edible fraction, can grow within an expanded range of salinity, are disease resistant, amenable for mass culture and possess many other improvements. But all these can materials only if more basic information is available on the physiology, biochemistry and genetics of cultured species and on refining the existing biotechnological techniques.

By far the importance of the applications of genetic engineering techniques is greatest to marine biotechnological field. It will also provide an untapped gene pool representing transport system for minerals, metal concentration, novel photosynthetic systems, marine pheromones, *i.e.*, communicator substances produced by marine organisms as well as the hydrogen sulphide utilizing and microbially mediated ecosystems. With the advent of the tools of genetic engineering, the potential of the sea as a significant source of protein feed can be assessed. The stock assessment and the migration of fish can be known so that we will know when and where to fish at a particular period. The management and stock breeding can be made with more precision using these biotechnologies for fish and shellfish in a more profitable way so as to provide feed and returns for our country.

References

Morris, H. Baslow, 1977. Marine Pharamacology. Robert, E. Krieger Publishing Co., Inc, New York. pp. 327.

Heinz, A. Hoppe and Tore LeVring, 1982. Marine algae in pharmaceutical science. Vol.2. Walter de Gruyter, Berlin. pp. 309.

Halstead, B.W. 1988. Poisonous and Venomous marine animals of the world. The Darwin Press, Inc, Princeton, N.J. U.S.A pp. 650.

Das, P., 1991. Genetics in enhancing aquaculture productivity. In Aquaculture Productivity (eds Sinha, V.R.P., H.C. Srivastava), Oxford and IBH Publishing Co., New Delhi. pp. 623-627.

Mary-Frances Thompson, Rachakonda Sarojini, Rachakonda Nagabhushanam, 1991. Bioactive compounds from marine organisms. Oxford and IBH Publishing Co.Pvt.Ltd. New Delhi. pp. 410.

James, P.S.B.R. and A.G. Ponniah, 1991. Concepts in marine biotechnology and their application for enhancing aquaculture productivity. In Aquaculture Productivity (Eds. Sinha, V.R.P and H.C. Srivastava), Oxford and IBH Publishing Co. New Delhi, pp. 559-605.

Anthony, T.T., 1993. Toxin related diseases. Oxford and IBH Publishing Co.Pvt.Ltd., New Delhi pp. 543.

Douglas Tave, 1993. Genetics for fish hatchery managers. Van Nostrand Reinhold, New York, pp. 415.

Ninawe,A.S., 1998. Biotechnological and Ecological approaches for conservation of fish species. In Fish gentics and biodiversity conservation, (Eds Ponniah, A.G., P.Das and S.R.Verma) NATCON, Pub No5, Muzaffarnagar, pp. 343-349.

Ponniah, A.G., P.Das, and S.R. Verma, 1988. Fish genetics and Biodiversity conservation. Nature conservators, Muzaffarnagar, India. pp. 474.

Chapter 3

Ornamental Fish in Applied Biotech Research

Saroj K. Swain, Snehasish Mishra, N. Rajesh and Ambekar E. Eknath

Central Institute of Freshwater Aquaculture, (ICAR) Kausalyaganga, Bhubaneswar – 751 002

Fish and particularly ornamental fish have umpteen uses in biotechnological research and are regarded as good scientific models. Ornamental fishes are considered as models for vertebrate research as they can be cultured in small aquarium in addition to high fecundity and external fertilisation as also developmental biology. In contrast, other well established test animals such as the rats, rabbits, guinea pigs and other such mammalian conventional test organisms have low fecundity, development is internal and lengthier developmental time as compared to few hours in fish. Once scientists genetically engineer the somatic cells or eggs and/or sperm, it must be inserted back into the mammals. However, it can be observed that it is *in vitro* in ornamental fish. As the ornamental fish are more sensitive to chemicals and pollution, it is a better choice for researchers for the study of ecotoxicity. The various fields where ornamental fish has a role as biotechnological tool: bioassay, pharmacotoxicology, eco-toxicology, toxicogenomics, toxicoproteomics, pathology, parasitology (host-pathogen interactions including *in vitro* interactive-enzymology), proteomics, genomics, phylogenetics, agriculture, germplasm conservation (cryopreservation), molecular characterisation to probiotics to elaborate a few state-of-art research *modus operandi*.

Ornamental Fish Species for Mass-Scale Production

Important egg-layers like barbs, rasboras, goldfish, tetras, danios, bettas and gouramis and major livebearers like guppies, platies, mollies and swordtails are good species as biotechnological research tools. Barbs are the most important group and most species of the group are known to have originated from India, *viz.*, rosy barb (*Puntius conchonius*), melon barb (*P. fasciatus fasciatus*), aruli barb (*P. arulius*) etc. Among the exotic barbs, tiger barb (*P. tetrazona*) and its aquarium-developed strains like green-tiger and red-tiger barbs are fascinating in modern days. The major species of the group danios include zebra danio (*Brachidanio rerio*), pearl danio (*B. albalineatus*), giant danio (*Danio aequipinnatus*), turquoise danio (*D. devario*), and malabar danio (*D. malabaricus*). Zebra danio is one of the typical examples of ornamental fish species of Indian origin, which can be bred and reared easily. Among rasboras, the slender rasbora (*Rasbora daniconius*), glowlight rasbora (*R. pauciperforata*), harlequin rasbora (*R. heteromorpha*) and scissors tail (*R. trilineata*) are important ones. Gold fish, *Carassius auratus*, is a commonly available fish, preferred by the hobbyists because of its attractive colouration ranging from pure gold to red, orange, black and albino. Some of the common varieties of gold fish available are comet, lion-head, oranda (a modification of lion head), fringe tail, veil tail, fantail, shubunkin, telescopic eye, etc. These fishes grow up to 20 cm in length, but start breeding when they are only 6 cm long. Many aquarium bred strains of ornamental koi, *Cyprinus carpio* (var. koi) can be bred easily by hormonal injections. Its origin is from East Asia but strains like orenji ogon, kin matsuba, kohaku and shiro-bekoo are being cultured worldwide.

The tetras are small fishes of 3-8 cm long; majority of which has originated from South America. The most common species of the group are black widow tetra (*Gymnocorymbus ternetzi*), serpae tetra (*Hyphessobrycon sarpae*), lemon tetra (*H. pulchripinnis*), flame tetra (*H. flammeus*), neon tetra (*Paracheriodon innesi*), cardinal tetra (*P. axelrodi*), black neon tetra (*Hyphessobrycon herbertaxelrodi*) and pretty tetra (*Hemigrammus pulcher*). The species, *Betta splendens* commonly called as Siamese fighting fish occurs in varied colours like green, red, blue, albino and sometimes with a combination of two or three shades. The attractive colour and hardiness of the species are the characters of the species for its wide adoption by the hobbyists. The males are brightly coloured with beautifully spread-over fins. They show aggressive behaviour, only when other males are present. Angel fish, *Pterophyllum altum* and *P. scalare* are important candidate species widely preferred for aquaria, with different developed varieties such as black, veil tail, gold and black marble, platinum, gold pearl, pink ghost and albino. The filamentous lower fins and their compressed body shape with their elegant movements are pleasant to look at. Among gouramies, three spot gourami (*Trichogaster trichopterus*), pearl gourami, (*T. leen*), moon light gourami (*T. microlepis*), snakeskin gourami (*T. pectorails*), dwarf gourami (*Colisa lalia*), honey gourami (*C. sota*), striped gourami (*C. fasciata*), noble gourami (*Ctenops nobilis*) and kissing gourami (*Holostoma temmincki*) are the important species. The bettas and gouramies are most popular among nest-builders, characterised by possession of accessory respiratory organs, thus making the species hardier. Among catfishes the exotic variety of sucker mouth catfish (*Hypostomus multiradiatus*) that originated from the South and Central America is very popular among aquarists.

Sexing the Fish

Determining the sex of a fish is an important aspect. Most fish can be classified as sexually dimorphic or sexually isomorphic. In sexually dimorphic species, the sexes can be easily distinguished by size, shape, colour and fin pattern. Like all other animals of animal kingdom males are more colorful, larger, and have more elaborate finnage. Among the more brilliant outstanding of sexual dimorphism can be found in cichlids, killifishes, barbs and Livebearers. In sexually isomorphic species, like angel fish, it is generally difficult to identify sexual differences. But an experienced aquarist can identify the sex by virtue of experience or working with that particular fish for a long time. Often, the only way to distinguish between the sexes is the shape of the genital papilla, which is only visible during spawning period. In some isomorphic species, the males are slightly larger and the females are slightly oval in the belly.

Selection of the Right Brooder

Once the sexes have been distinguished, a suitable pair or spawning group can be selected. There are several important traits to seek in choosing the brood fish. Select the fish that shows good markings and colour that should produce attractive young ones. Always use mature, healthy fish for spawning because unhealthy fish, if they spawn, may produce unhealthy or deformed hatchlings. Ensure that the pair is well compatible. Many species cannot be put together in a breeding tank and expected to get along and produce young. In many cichlids, pairs form only after a group has been raised together for months. In certain species like gouramies, during spawning process, if the male is ready and female is not, in that case male may kill the female. This may be due to incompatibility among the sexes. Try to avoid crossing between the different strains or color forms because the young are often unattractive. Always ensure that the pairs are from the same species because usually hybrids are sterile.

Conditioning the Brooders

Before placing the parent fish together for spawning, they should be conditioned through best feeding strategies with a variety of live feeds to get them in excellent matured condition for spawning. The live feeds such as tubifex, blood worm, zooplanktons, etc. which not only gives the good growth but also triggers the spawning process.

Breeding Strategies and Spawning Environment

Some of the ornamental fish species readily spawn in the aquarium or cement tanks, the eggs or hatchlings often do not survive because of predatory nature of the parents or other fish. Sometimes the mortality occurs due to unfavorable, polluted water conditions. It is always better to breed the fish in a separate spawning tank. The spawning tank may be fitted with a thermostat (if necessary), a sponge filter and good aeration. Depending on the spawning method, the spawning tank can be set up in a number of different ways based on the spawning behaviour of the fish.

Oviparous (Egg Layers)

Most of the aquarium fish are egg-layers with external fertilisation. Egg-layers can be divided into five groups: egg-scatterers, egg-depositors, egg-burriers, mouth-brooders and nest-builders.

Egg-Scatterers:

Zebra fish (*Zebra danio*) is considered as egg scatterers,lay non-adhesive eggs. The larger ones are grouped under the genus *Danio* and other smaller varieties are under genus *Brachidanio*. The important varieties of danios include the giant danio (*Danio aequipinnatus*), pearl danio (*Brachidanio albalineatus*) and zebra danio (*Brachidanio rario*). Since the egg scatterers often eat their own eggs, the spawning tank has to be set in such a way that the eggs fall out of the reach of spawning parents. These species do not show parental care and eat their own eggs.

For breeding such fishes the male and female ratio should be maintained at 1:1 or 2:1. 40-50 l capacity tank is sufficient for spawning for most eggs scatterers. The female is introduced in the breeding tank one day earlier than the males. But it is very difficult to know whether the female has already laid the eggs because of smaller size of eggs and remain hidden behind the pebbles. Sometimes the brood fishes are kept in a mesh net from where the eggs drop in to the bottom. If the temperature is favourable, the eggs require two-three days hatching time. As soon as the tiny hatchlings seen in the aquarium tanks the parents are to be removed. The hatchlings take 2-3 days to absorb their yellow yolk sac. Then they are fed with infusorians for 4 days followed by rotifers and smaller zooplankton for a week, after which they can be provided powdered formulated feed along with zooplankton.

The egg scatters laying adhesive eggs are mostly the Gold fishes, *Carassius auratus*. The common gold fish varieties are common gold fish, fringe tail, lion head, oranda, comet, shubunkin, telescopic eye, veil tail, red cap, etc. When secondary sexual characters appears (by observing the maturity condition), the male and female gold fish are selected and kept in glass tanks or ferro-cement tanks. Since gold fish eggs are sticky in nature, they require some surface to adhere their eggs. Various types of natural submerged aquatic plants like *Hydrilla* can be used after treatment for this purpose. Artificial nests can be prepared by making split plastic ropes with one end tied or burnt to make it blunt. Even polythene strips have been found suitable for the purpose.

The female and male in the ratio of 1:2 are released into breeding tank during late evening hours. The male chases the female, presses its operculum against female's abdomen and fertilizes the eggs while swimming beside her. Egg laying usually takes place within 6-12 hrs of releasing the broods. After spawning, the nest is transferred to a different container, or alternatively, the parent fish is transferred from the breeding tank. If this is not done, the parents are most likely to eat away the eggs to compensate the post-spawning loss of energy.

Generally, the female lays about 2000-3000 eggs. Healthy eggs are golden transparent in colour and the unfertilized eggs will remain opaque and continue to remain as such with arrested growth. These dead eggs become pale white and hair like aquatic fungus would grow on all the sides, giving it the appearance of a small powder-puff. Under ideal condition, within three days, the eggs hatch out with a hatching rate of 80-90 per cent when the young larvae start to float, the nest materials are taken out from the tank. Generally, the tiny hatchlings remain clinging to the nest, so precaution has to be taken while transferring the nest from the breeding pool.

Egg-Depositors

These species deposit their eggs on a substrate (tank glass, wood, rocks and plants). Egg depositors usually lay less egg than egg-scatterers, although the eggs are larger. Depending on the type of egg depositor, the tank should be furnished differently. For those egg-depositors that care for their young, the parents can remain in the tank after spawning.

Substrate spawners, depending on the species, should be given a tank furnished glass panes, broad-leafed plants, or flat stones for spawning sites. Some species such as Discus and Angelfish prefer vertical surfaces. From a group of such fishes a pair is selected by providing a slanting serrated glass surface plate, where a suitable pair comes near the plate for cleaning the surface for laying eggs. At that time the pair is carefully removed for breeding in a separate tank. For cavity spawners like badis fish, flower pots turned on their side, coconut shells, and rocky caves are suitable spawning sites. The tank should be furnished with either live or plastic plants to give the fish a sense of security.

Egg-depositors that do not care for their young ones should be given a tank furnished with fine and broad-leafed plants. After spawning the parents or plants with the eggs are to be removed. If the plants containing eggs are removed, new plants should be placed in the tank for future spawning. The very common ones are *Rosbora daniconius* (Slender rasbora). They require soft, slightly acidic (pH 5.5-6.5) and temperature at 28°C. After conditioning male and female they are placed in a tank planted with flat leaved plants. Rasboras prefers peace and quiet environment for breeding with low lighting conditions. The male and female brooders are placed together in breeding tank. Once spawning has occurred, as indicated by the slimness of the female fish, remove both parents from the breeding tank. The eggs laid on the underside of the flat levels will hatch after 25-30 hours. Generally from a larger female up to 250 eggs are produced and resultant hatchlings become free swimming after 3-4 days. At this stage the tiny hatchlings should fed infusorians and newly hatched brine shrimp. As they grow bigger they should be fed with zooplankton. They can also bred by another method by putting the gravid brood in a mesh net through which eggs are dropped in to the bottom and hatching takes place.

Egg-Burriers

These species usually inhabit waters that dry up at some time of the year. The majority of egg burriers are Killifish which lay their eggs in mud. The mature parents lay their eggs before dying when the water dries up. The eggs remain in a dormant stage until the fresh rains and stimulate hatching. These fishes are *Aplocheilius lineatus, A. panchax, A. dayi and A parvus.* A peat-moss substrate is one of the best substrates for egg-burying species as mentioned by some of the authors. In order to initiate hatching, the stored peat can be immersed in soft water.

Nest-Builders

Nest builders build the nest for their eggs. The nest is usually in the form of bubble-nest formed with plant debris and saliva-coated bubbles (labyrinth fish), or an excavated pit in the substrate (cichlids). Nest builders practice brood care. Nest-builders should be provided with a wide leafy material with which to build their nests and the tank should have no water current to disturb the nest. Species that build nests are coming under Gouramies (*Colisa lalia, C. fasciata)* and fighter fish (*Betta splendens*).

For breeding such fishes, the sexes are kept separately for few weeks. As the abdomen of female becomes grossly distended with eggs it is transferred to a smaller breeding tank containing few floating plants with water level of 5-6" at a temperature of 28-30°C. After one or two days a good male is introduced in the breeding tank. A transparent perforated plastic sheet or a glass is covered over the tank to keep the humidity and temperature at high level and help to maintain the bubble nest in good condition. The male soon begins building a bubble nest. This is possible by taking a large gulp of air at the water surface and converting it into many smaller bubbles that are passed into gill chamber and coated with an anti-burst agent before release. During and after making the nest, the male displays to the female which usually ends with both fish embracing near the nest resulting deposition of a large numbers of eggs in the nest. After making the bubble nest if the female do not lay eggs then the male become very aggressive and may kill the female. After breeding, the female is removed. The male guards the eggs which remain attached to the floating bubble nest. Within 24 hours, hatching takes place. The moment, the fry begin leaving the nest, the male is also removed from the tank. After 36 hours, the young ones remain in free swimming stage, they are provided infusorians as starter feed. After a week the fry starts taking newly small zooplankton. During this stage fry requires intensive feeding. Subsequently when they grow a little bigger they can be stocked in larger cement tanks for further growth.

Mouth-Brooders

Are species that carry their eggs or larvae in their mouth. Mouth brooders can be devided into ovophiles and larvophiles group. Ovophile or egg-loving mouth-brooders lay their eggs in a pit, which are sucked up into the mouth of the female. The small number of large eggs hatches in the mother's mouth, and the fry remain there for a period of time. Fertilization often occurs with the help of egg-spots, which are colorful spots on the anal fin of the male. When the female sees these spots, she tries to pick up the egg-spots, but instead gets a mouthful of sperm, fertilizing the eggs in her mouth. Many cichlids and some labyrinth fish are ovophile mouth brooders.

Larvophiles or larvae-loving mouth-brooders lay their eggs on a substrate and guard them until the eggs hatch. After hatching, the female picks up the fry and keeps them in her mouth. When the fry can fend for themselves, they are released. Ovophile mouth-brooders can be bred in the main aquarium because the eggs are protected in the mouth cavity. However, it is better to separate mouth-brooders with eggs because of their potentially aggressive behavior. There are no special breeding tank requirements other than the usual tank set-up for the species. Larvophile mouth-brooders should be placed in a breeding tank because the eggs are not protected in the mouth, but laid on a surface where they are open to predators. For mouth brooders open cement cisterns of 2 feet water depth with planted long leafed plants are generally used for commercial production.

Ovo-Viviparous (Livebearers)

Livebearers are fish that bear live young ones. They are ovoviviparous in nature, where the eggs form and hatch within the female before birth. Livebearers are often prolific, easily bred species. Breeding of most of the livebearers is relatively easy. They are mostly molly, platy, swordtail and platy. Development of young ones takes place inside the female body and they released after about four weeks. The species of livebearers include guppy (*Poecilia reticulata*), black molly (*Poecilia sphenops*), swordtail (*Xiphophorus helleri*) and platy (*X. maculatus*). The number of babies produced by a livebearer is normally 50-60 only, though some larger swordtails and guppies may produce as many as 100 nos. If the livebearers are fed properly with natural feed and supplemented with better artificial feed, the mother produces more than 100 young ones. Soon after the baby comes out from the mother, they have to be separated out to reduce mortality caused by predation. Small livebearers can be bred in breeding traps where the newborns come out of the reach of the brood mother. However, a more traditionally preferable set-up is a separate, densely planted tank where only gravid females are kept. As the young ones are visible in the water surface, they are to be removed and fed the powdered feed and zooplanktons. As soon as all the young are born, the mother is to be removed and fed suitably.

Ornamental Fishes as Biotechnological Tool in Disease Study

Zebra fish, *Brachydanio rerio* is used as a major test organism in biotechnology. Its transparent embryos like mammals are ideal to study the link between development and immunity. Zebra fish embryos already possess immune-competent macrophages after one day of development. Genome analysis shows that 90 per cent of all human genes have orthologs in zebrafish. In addition to *in vivo* studies, we use various zebrafish and human cell lines for signal transduction research. Beginning with the pioneering work of Streisinger and colleagues, the zebrafish was envisioned as an excellent model system for studying complex biological system (Amatruda *et al.*, 2002).

Zebrafish models were made for human development and disease study (EU 6th framework). In this European project micro-arrays and proteomics are being used to aid the study of signalling pathways induced by glycan molecules that are involved in the communication between cells in relation to pathogen recognition, development and cancer. The signalling functions of glycans that represent ligands of the Toll-like receptors, such as hyaluronan oligosaccharides were studied.

Ornamental Fishes as Biotechnological Tool in Etioloy

Comparative characterisation, pathogenicity, biochemical, genomic and serological studies showed nodavirus infection in various isolates of diseased guppy, *Poicelia reticulata* (Hegde *et al.*, 2003). The biotechnological tools/methods involved in the study are worth-mentioning. The nucleotides of various strains were compared. Western blot analysis using rabbit antisera raised against the nodavirus from marine fish, *Epinephelus tauvina* confirmed its antigenic similarity to the marine nodavirus isolate. Asymptomatic infection in guppy fry was observed following experimental infection with this virus and the marine nodavirus isolate (Singapore strain) implying the spread of virus from marine fish to freshwater fish. Thus, by use of above methods first nodavirus infection in freshwater fish was confirmed.

Nightlights for the Aquarium Used as Research Tools

Genetic tinkering has resulted in fluorescent fish and a boost for Taiwanese ornamental fish dealers. These can also be used as GM house pets and known as Night Pearl. The technology for the breakthrough was based on the research of Tsai Huai-jen of the National Taiwan University (NTU). In 2001, he announced creation of fluorescent fish and saw the potential to sell them as aquarium fish. The genetically-modified trait of spectral glow by the otherwise ordinary-looking Japanese medaka and zebra fish made these livelier for the aquarium tank. However, the environmental concerns may not be overlooked as there is a chance of escape to nature causing damage to nature.

Ornamental Fish as 'Bioreactors'

At present the most promising application of transgenic technology in animals is to use transgenic animals as a 'bioreactor' to produce useful proteins and compounds in medicine and other areas. This is also known as 'gene farming' where gene of other animal or species can be inserted within a specific tissue so that it produces useful substances. Current biotechnological research efforts are not limited to improving farm productivity. In an attempt to identify new sources for production of pharmaceutical substances, scientists have engineered tilapia with a human gene that makes the fish capable of producing a compound essential to the production of Factor VII, a human blood-clotting protein. Scientists have tried to improve cold tolerance of goldfish by engineering hsp (heat-shock protein) genes so that it thrive cutting across global agro-climatic zones, than the wild counterpart (Hwang *et al.*, 2004).

The advantages of the transgenic fish bioreactor include the speed of generation of transgenic fish, low cost, and low risk of transmitting animal pathogens, etc. To increase the yield of the fish muscle bioreactor system, fast-growing and large size farm fish such as carp, tilapia, catfish, salmon and rainbow trout may be used. For all these species, the transgenic technology is well developed (Gong and Hew, 1995).

Ornamental Fishes as Tools in Environmental Biotechnology

Ornamental fish are being used as test organism for toxicity study through bioassay. This focus includes the sub-disciplines of neurotoxicology, immunotoxicology and environmental toxicology. Ornamental fishes are exposed to increasing problem of aquatic pollution and many species can only be saved through breeding programs. Several studies have documented heavy metal toxicity on growth uptake and elimination of metals (Cuvin and Furness, 1988), reproduction (Kaviraj, 1983; Nicoletto and Hendricks, 1988) and metal removal by biochemical agents (James *et al.*, 2000) in cultivable fishes. The findings of copper toxicity on aquarium fish could be applicable to cultivable fishes too. The toxic impact of copper on growth and reproductive performance in sword tail, *Xiphophorus helleri* has been reported. The major effects were vertebral deformity, low reproductive potential and growth reduction. A semi-static bioassay was used to study acute toxicity of endosulfan to zebrafish (*Brachydanio rerio*) and yellow tetra (*Hyphessobrycon bifasciatus*) which included behavioral changes like hyperactivity, erratic swimming and convulsions when exposed to highest toxicity levels.

Ornamental Fish as Tools in Molecular Biology

Molecular-level manipulation by employing basic principles of molecular genetics and allied techniques could find solutions for increasing the overall fish production. The molecular level studies could be used in several areas such as population genetics, brood stock development, fish health management, control of reproductive cycles, transgenics, genetic diversity and conservation, fish genomics etc. In fisheries biology, molecular markers can be used as valuable tools in systematics, phylogenetics, brood stock management, population structure, species identification, gene mapping, selective breeding, QTL mapping, legal applications and many others. GM fish offer new potential for increased production and value addition of cultured organisms. This technology allows the introduction of new traits or the improvement of the old ones in a way that is almost impossible to achieve with conventional breeding methods. Examples of genes with commercial potential are those which control growth, disease, cold tolerance, sexual maturation and meat quality/ preservation.

Protein Markers in Phylogenetic Study of Ornamental Fishes

Genetic variation will ultimately tend to proteomic polymorphism within limited time span and morphological differences in broad spectrum of time frame. The similarity in protein banding pattern between Goldfish, *Carassius auratus* and Koi Carp, *Cyprinus carpio* may be attributed to the biochemical and metabolic correspondence among the fishes. Although differences among species in enzyme maximal activity or concentration are often interpreted as adaptive and important for regulating metabolism, these differences may simply reflect phylogenetic divergence. The results indicate similar metabolic requirements of these enzymes in the particular organs. Various biotechnologically useful enzymes and proteins can be compared and characterised in this phylogenetic study.

Embryonic Stem Cell Research using Ornamental Fish

The cryopreservation protocol for embryonic stem (ES)-like cells of a tiny freshwater fish Leopard danio (*Brachydanio frankei*) is possible. Embryonic stem (ES)-

like cells derived from blastomeres of the early blastulae stage of the developing embryo were cultured *in vitro* at CIFA stem cell laboratory (Routray *et al.*, 2010). Embryonic stem cells in general possess a self-renewal capacity and developmental potential to differentiate into any cell type of an organism and can be regulated and maintained *in vitro*. These cells provide a good *in vitro* model system to understand the molecular mechanism of pluripotency, conservation and basic biomedical researches.

Ornamental Fish in Germplasm Conservation

Fish variety gets hybridized with undesirable morphologically similar variety. This hybridisation, either naturally, or directed by man, has widely propitiated the contamination of cultured stocks. Thus, conservation of ornamental fish germplasm is the need of the hour. For the study of phylogenetic variation in fishes, various scientists have differentiated them on the basis of morphological, molecular and biogeographical patterns.

Ornamental Fish in Molecular Characterisation and Phylogeny

The main sources of markers are mitochondrial DNA (mtDNA), minisatellites, microsatellites, introns, and anonymous nuclear sequences assayed using highly specific PCR primers or DNA probes including RAPD technique. Gold fish is considered as one of the commercial ornamental fish in the trade (Bandyopadhyay, 2004). Goldfish being a Cyprinid could be used as an ideal model for biotechnological studies in larval and juvenile cyprinids. It has huge commercial value both in foreign and domestic market. Due to its commercial importance and high reproductive ability, it is necessary to develop an improved strain. Some efforts have been taken to assess the fish germplasm (Bardakci and Skibinski, 1994). A phylogenetic profile is a pattern of presence or absence of a particular gene or protein across a set of organisms. Proteins that function in the same cellular context frequently have similar phylogenetic profiles which have proven to be helpful in the elucidation of the evolutionary history of biochemical pathways or protein machineries. The fast growing number of completely sequenced genomes continuously enhances the resolution of phylogenetic profiles. The studies shows that DNA markers are useful tools for estimating inbreeding depression and heterosis in guppy (*Poecilia reticulata*) breeding (Shikano and Taniguchi, 2003). Protein as well as DNA-RAPD studies showed that Goldfish is phylogenetically more related to koi than any other ornamental fishes among goldfish, koi, texas and gourami. Phylogenetically, Gourami was most distant to goldfish (Malik, 2006).

Ornamental Fish as Tool for 'Probiotics' Research

A termed coined and first used in aquaculture about two-decades before by Jory *et al.* (1986), probiotics (pro-favour; bios-living beings) unlike antibiotics refer to organisms that favour the growth of a target organism. 'Probiotics' generally includes bacteria, cyanobacteria, micro-algae and fungi etc. referring to all those microbes which could improve general water quality inhibiting the water-borne pathogens, thereby increasing productivity. Probiotics increase the population of fish-feed organisms; improve the nutritional levels of aquacultural animals and their immunity to pathogens. Sugita and Shibuga (1996) isolated bacteria from the intestine of 7 kinds of freshwater cultured fish and reported their antibacterial abilities. They concluded that the presence of intestinal bacteria protected the fish against the infection by pathogens. Researchers at Heriot-Watt University have developed a range of probiotic micro-organisms which show protection against fish diseases such as furunculosis, enteric redmouth disease, vibriosis and winter ulcer disease. In addition these organisms provide a dietary supplement to enhance the growth, nutrient utilization and overall health of carp species (Swain *et al.*, 1996). All these studies can easily be undertaken in any ornamental fishes due to the advantage culturing in confined environment without polluting the nature.

Ornamental Fish as Tool in Reproductive Biology

The Reproductive Behaviour of *Asian Arowana*

Pioneering research collaboration is on with Qian Hu Fish Farm on the reproductive biology of a primitive teleost Asian arowana (*Scleropages formosus, Osteoglossidae*) or dragonfish. Asian arowanas are interesting for basic research due to their unusual reproductive strategy (a few eggs of huge size protected by mouth brooding) and they command an extremely high price in the ornamental fish trade as these are regarded as most favorite omen in '*Feng Sui*'. Their partner selection process is being studied by using genotyping with polymorphic DNA markers isolated in the lab and trying to develop methods to predict mate choice on the basis of genotypes.

Genomic Analysis of Teleost Sex Chromosomes

The majority of the 24 thousand fish species do not seem to have highly differentiated sex chromosomes and those that do seem to show tremendous variability. Differences among male and female genomes are being searched by PCR-based methods during the past ten years and DNA markers associated with sex from three different teleost species (African catfish, Asian arowana and Turbot) have been successfully isolated. Sex-associated DNA markers might be used for increasing the density of genetic linkage map of sex chromosomes. The comparative analysis of sex chromosomes will help to understand the secrets of their evolution. Sex-specific DNA markers allow developing "molecular sexing methods", which are useful for improving the aquaculture production of farm fish species with differential growth rate and maturation time between the two sexes.

Ornamental Fish as Tool in Ontogeny

Transgenics is becoming increasingly popular as a powerful experimental tool in developmental biology, notably in zebrafish (*Brachydanio rerio*) and medaka (*Oryzias latipes*). From late 1980s to late 1990s, the research in this area is basically for technology development. With the demonstration of faithful expression of the *GFP* (*green fluorescent protein*) living color reporter gene under homologous tissue specific promoters (Higashijima *et al.*, 1997; Ju *et al.*, 1999), the power of the *GFP* transgenic fish has been quickly recognised. Now, the *GFP* transgenic fish system is actively used in gene expression patterns study, tissue/organ development, tissue-specific promoters/enhancers, cell lineage and migration, upstream regulatory genes, mutagenesis screening and characterisation, promoter/enhancer, chimeric embryos and fish cloning by nuclear transplantation, etc. (Gong *et al.*, 2002, 2004).

Comparative Genomic Study of Gonad Differentiation in Cyprinid Teleost

Genes with conserved regulatory roles in the gonad differentiation of common carp (*Cyprinus carpio L., Cyprinidae*) and zebrafish are being searched in the genomic library. By using testis samples from all-male common carp populations generated through androgenesis, cross-species hybridization experiments with the microarray are being performed. Candidate genes with putative role in the gonad differentiation in teleosts have been studied.

Ornamental Fish for Hybridisation Aberration Studies–Mosaic Fish

In aquaculture, the most successful application of the transgenics should be the generation of growth enhanced *GH* transgenic fish. It's observed that cytoplasm affects the number of vertebrae in Carp-goldfish clones. Although the initial rate of success in producing carp-goldfish clones is low, it was believed that cross-species transplantation will lead to improved understanding of the contributions of the nucleus and egg cytoplasm to the growth and development of vertebrates (Sun *et al.*, 2005). Ornamental fish can also be used as biotechnological tool for developing mosaic fish. Colour gene was derived by two tissue specific promoters, one for skin cells because of the exterior location and the other for muscle cells because of the large mass. These promoters on ligation to the jellyfish *GFP* gene generated stable transgenic zebrafish lines using both chimeric genes. Another colourful transgenic zebra-fish was developed, using the same muscle specific *mylz2* promoter with RFP (red fluorescent protein) and YFP (yellow fluorescent protein) genes (Gong *et al.*, 2003). These glow-fishes are currently being marketed in the US.

Future Prospects and Concerns

Using biotechnological techniques, including molecular and recombinant technology, scientists study the growth and development of ornamental fish to understand the biological basis of traits such as growth rate, disease resistance or resistance to destructive environmental conditions. Researchers are using biotechnology to identify and combine valuable traits in distant-parental fish to increase productivity and improve product quality. The traits scientists and companies are investigating for possible incorporation into several organisms include increased production of natural fish growth factors and the natural defense compounds organisms use to fight microbial infections. Biotechnology is also improving productivity through the development of feed additives, vaccines and other pharmaceutical agents.

Ornamental fishes are small, easy to handle, can be studied and maintained in the aquarium as test animals. An aquarium makes an ideal ecosystem to study interaction of fishes with their surroundings. In environmental biomonitoring, though all the animals are affected by pollution but only fishes are taken as model as these are susceptible to diseases and sensitive to pollution.

The benefits from such transgenic fish are obvious, *e.g.*, increase of yield, shortening of the production time, decrease of operation cost, potentially lowering the market price for consumers' benefit, etc. There is no indication that fluorescent transgenic zebrafish pose a higher risk than the wild type fish to ecosystem/ environment. A commitment to research and to development in ornamental biotechnology will respond to the critical needs of society by developing novel drugs for treating diseases, providing new techniques to monitor and assess, restore, protect and monitor environments, ensure sustainable and safe aquaculture and fisheries, discovering new types of composite materials, biopolymers and enzymes for industry.

A major concern is the marketing of transgenic fish, for environmental reasons (not limited to GM fish, *e.g.*, exotic species etc.), and ethical concerns. Viability of transgenic fish is no better than that of the wild type, lower fecundity and fertility. Though the ethical committees have put many restrictions and legislations behind the use of animals like guinea pigs, rats etc., they have not kept any such limitations on the use of fishes. Hence, these can be safely used as research models. In recent years, scientists have applied genetic engineering for a wide variety of purposes to a host of animals traditionally used as feed sources, including cows, pigs and fish.

Conclusion

Ornamental fish are the most valuable fisheries commodity in the world today in terms of returns per unit area. Further, owing to the fact that these are such suitable specimen from conventional to the most modern type of biological research, some of the concerns that need proper and immediate address include, survival of existing species; maintaining the potential of species for continual evolution; utilitarian value; collection of baseline data on ecosystem, socioeconomic and diversity of the population in order to analyze the impact of the ornamental fish trade on social and natural environments; expanding their biodiversity and genetic characterization studies. The 'ranching' works as carried out by many Governmental and Non-governmental bodies may further be encouraged. To conclude, ornamental fishes could prove beneficial in disease study, bioassay, toxicology study, pathogen study, proteomics and genomics, host-pathogen interactions in phylogenetics study, *in vitro* cellulase culture (economical cellulase extraction) study, agriculture, germplasm conservation, molecular characterisation and phylogeny study, as also probiotics and other such multi-disciplinary applied research.

Acknowledgements

The first three authors are grateful to Dr. AE Eknath, Director, CIFA, Bhubaneswar for his continuous support and guidance.

References

Amatruda, J. F., Shepard, J. L., Stern, H. M and Zon, L. I. 2002.Zebrafish as a cancer model system, Cancer cell. 1:229-231 (3).

Bandyopadhyay, P., Swain, S. K., Mishra, S. 2004. Growth and diatary utilization in goldfish (*Carassiusauratus* Linn.) fed diets formulated with various local agro-produces. *Bioresources Technology* 96 (2005) 731-740.

Bardakci, F and Skibinski, F. O. D. 1994. Application of RAPD technique in tilapia fish: species and sub-species identification. *The genetically society of Great Britain.* 71: 117-123

Cuvin, M. L. A and Furness, R. W. 1988. Uptake the elimination of inorganic mercury and selenium by minnows *Phoxinusphoxinus, Aquatic taxology.* 13 (3): 205-216.

Gong, Z., Ju, B., Wang, X., He, J., Wan, H., Putter, M.S and Yan, T. 2002. Green fluorescent protein expression in germ-line transmitted transgenic zebrafish under a stratified epithelial promoter from keratin 8. Dev. Dyn. 223: 204-215.

Gong, Z., Zeng Z., Pan X., Wan H., Ju B., Liu X. and Zhan H. 2004 Applications of transgenic fish technology. J. Virol., 78(22): 12576-12590.

Gong, Z., Hew, C. L. 1995. Transgenic fish in aquaculture and development biology. *Curr Top del Biol.* 30: 177-214.

Hegde, A., The, H. C., Lam, T. J and Sin, Y. M. 2003. Nodavirus infection in freshwater ornamental fish, guppy, *Poiceliareticulata* – comparative characterization and pathogenicity studies, *Archives of Virology* 148(3):575-586.

Higashijima, S., Okamoto, H., Ueno, N., Hotta, Y and Eguchi, G. 1997. High-frequency generation of transgenic zebrafish which reliably express GFP in whole muscles or the whole body by using promoters of zebrafish origin *Dev. Biol.* 192:289-299.

Hwang, G., Muller, F., Rahman, M. A., Williams, D.W., Murdock, P. J., Pasi, K. J., Goldspink, G., Farahmand, H and Maclean, N. 2004. Fish as bioreactors: Transgene Expression of Human Coagulation Factor VII in Fish Embryos, *Journal of Marine Biotechnology*. 6:485-492.

James, R and Sampath, K and Selvamani, P. 2000. Effect of ion exchanging agent, zeolite on removal of copper in water and improvement of growth in *Oreochromis mossambicus* (Peters). *Asian Fish. Sci.* 13: 317-325.

Jory DE.1998. Use of probiotics in penaeid shrimp growout.*Aquaculture Magazine*,24: 62–67

Ju, B., Xu, Y., He, J., Liao, J., Yan, T., Hew, C. L., Lam, T. J and Gong, Z. Y. 1999. Faithful expression of green fluorescent protein (GFP) in transgenic zebrafish embryos under control of zebrafish gene promoters. *Dev. Genet.* 25(2):158-167.

Kaviraj, A. 1983. Effects of mercury on the behavior, survival, growth and reproduction of fish and on aquatic ecosystem. *Environment and Ecology*, 1(1): 4-9.

Malik, D. 2006. Phylogenetic profiling of some important ornamental fishes and their comparative cellulose assay. M.Sc. Dissertation. CCS University, UP, India: 76-87.

Nicoletto, P. F and Hendricks, A. C. 1988. Sexual differences in accumulation of mercury in four species of centrarchid fishes. *Can. J. Zool.* 66(4): 944-949.

Routray, P., Dash, C., Dash, S. N., Tripathy, S., Dhananjay, K. V., Swain, S. K., Swain, P and Guru. B. C. 2010. Cryopreservation of isolated blastomeres and embryonic stem-like cells of Leopard danio, *Brachydaniofrankei, Aquaculture Research*. Vol. 41, No. 4: 579-589

Shikano, T., Taniguchi, N. 2003. DNA markers for estimation of inbreeding depression and heterosis in the guppy *Poecilia reticulate,* Aquaculture *Research,* Blackwell Publishing Volume 34(7):905-911.

Sugita, H. Shibuya, K. 1996. Antibacterial abilities of intestinal bacteria in freshwater cultured fish. 145 (1/4):195-203.

Sun, Y., Chen, S., Wang, Y., Hu, W and Zhu, Z. 2005. Cytoplasm affects the number of Vertebrate in Carp-goldfish Clones. Society for the study of reproduction.

Swain, S. K., Rangacharyulu, P. V., Sarkar, S and Das, K. M. 1996. Effect of probiotic supplement on growth and nutrient utilization and carcass composition in mrigal fry. *J. Aqua.,* 4: 29-35.

Part II

Fermentation Biotechnology

Chapter 4

Fermentors and Fermentation Techniques: Principle, Protocol and Applications

S. Felix and P. Pradeepa

Fisheries Research and Extension Centre,
Tamil Nadu Fisheries University,
Madhavaram, Chennai – 51, T.N.

Fermentors or bioreactors is a device in which substrates of low value is utilized by living cells or enzymes to generate a product of higher value.

General Characteristics of Fermentors

Fermentors are meant for the best growth and biosynthesis conditions for microbial cultures and for easy manipulation of all operations. For these purposes fermentor vessel must be made up of strong metal to withstand the pressure of large volume of medium. Material must not be corroded by fermentation product nor contribute toxic ion to the growth medium. Materials selected depends upon the type of fermentation.*E.g.* wooden tank is used for production of lactic acid. Fermentor vessels can be coated with materials like copper, stainless steel, iron and glass.

Provisions in the Fermentor

Fermentation process mainly uses pure cultures; to prevent the growth of contaminating microorganisms.

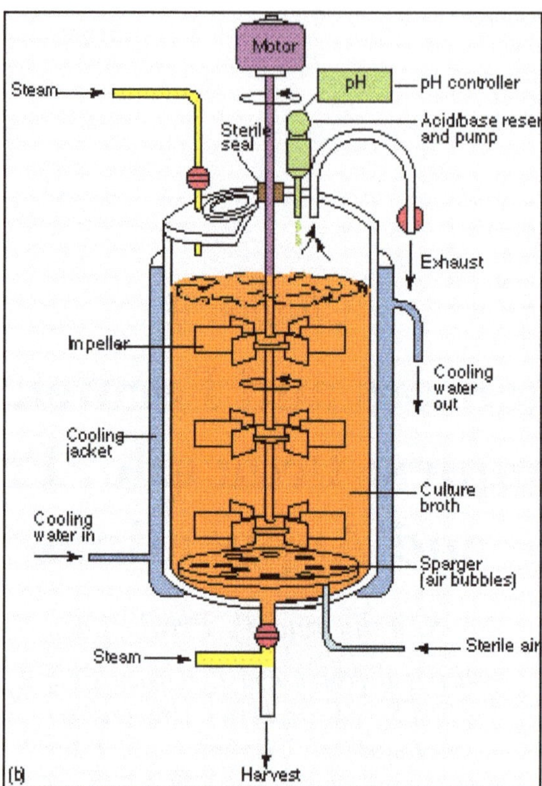

Figure 4.1: Design details of a Fermentor

1. Provision for Air Supply

For aerobic fermentation there should be provision for rapid incorporation of sterile air into the medium. The air should be supplied in such a way that oxygen in the air dissolves in the medium and will be available readily for microbes. Sparger helps for that purpose. Sparger bursts the large bubbles and leads to bursting and formation of smaller bubbles for easy dissolution.

2. Stirring device

A fermentor should be provided with a stirring device for the uniform distribution of air, nutrients and microbes. The device consists of a strong and straight shaft to which impellors are fitted. The impeller consists of circular discs to which blades are fitted. Different types of blades are available and used according to the requirements.

3. Baffles

It brings about transfer of turbulence to the fermentor wall. Four baffles are commonly installed with a width of 1 : 10 or 1 : 12.

4. Foam Control

The fermentor should be provided for the addition of antifoam agents. Antifoam agents lower the surface tension which decreases stability of the foam bubbles so that they burst. Two types of antifoam agents are used they are inert antifoams and antifoam agents made from crude organic material which is sterilized separately and added when needed. Only enough antifoam is added otherwise it will be toxic to microbes. Examples of crude antifoam are animal oil, vegetable oil, lard oil, corn oil and soybean oil.

5. Temperature Control

Temperature control is achieved by a water jacket around the vessel. Jacket is supplemented by internal coils to provide sufficient heat transfer surface.

6. pH Control

pH control is achieved by acid or alkali and controlled by an autotitrator which is connected to pH probe.

7. Addition Ports

Addition port is used for addition of culture inoculum and media requirements. In addition there should be proper inlet for introducing media and outlet for harvesting 20- 25 per cent fermentor volume should be left free called head space.

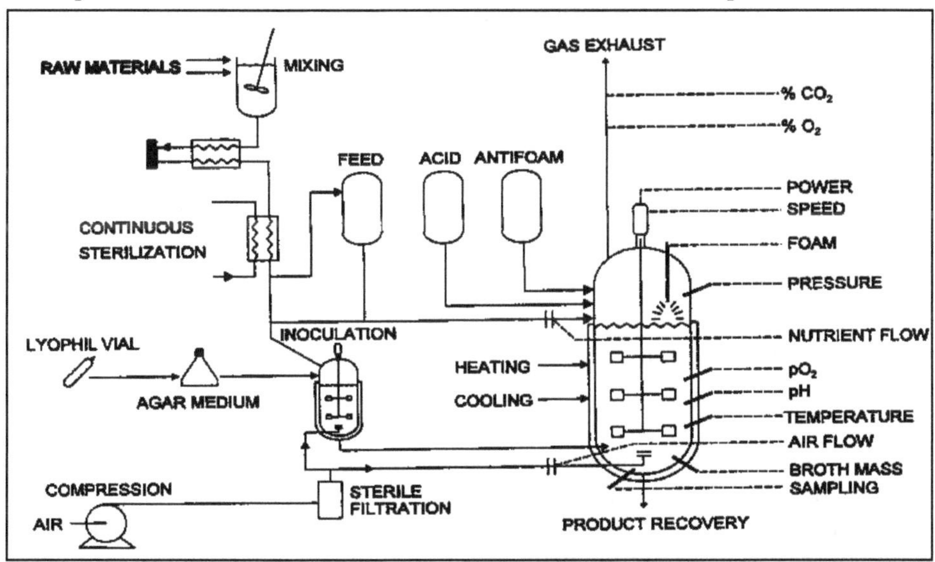

Figure 4.2: Flow Sheet of a Multipurpose Fermentor and its Auxiliary Equipment

Types of Fermentors

1. Simple Fermentors

i. Batch Fermentor

In the batch fermentation process the entire medium is removed from the fermentation tank. The tank is then thoroughly washed, cleaned and the new batch is

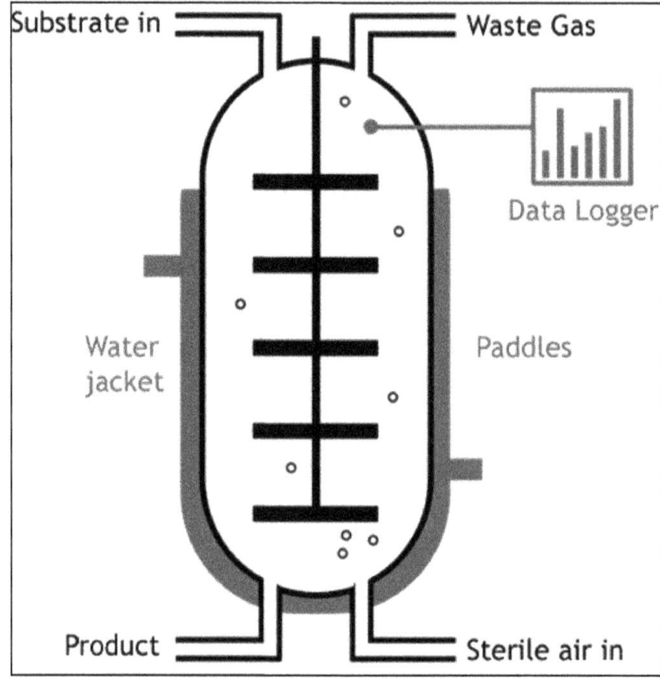

Figure 4.3: Batch Fermentor

started only there after. A tank of fermentor is filled with the prepared mash of raw materials to be fermented. The temperature and pH for microbial fermentation is properly adjusted, and occasionally nutritive supplements are added to the prepared mash. The mash is steam sterilized in a pure culture process. The inoculum of a pure culture is added to the fermentor, from a separate pure culture vessel. Fermentation proceeds, and after the proper time the contents of the fermentor, are taken out for further processing. The fermentor is cleaned and the process is repeated. Thus each fermentation is a discontinuous process divided into batches.

ii. Continuous Fermentor

In the continuous fermentation process, a part of the medium is removed at more or less regular intervals when the fermentation process is in force, and new medium is added therein. Thus, the process of fermentation continues non-stop. In continuous fermentation, the substrate is added to the fermentor continuously at a fixed rate. This maintains the organisms in the logarithmic growth phase. The fermentation products are taken out continuously. The design and arrangements for continuous fermentation are somewhat complex.

2. Fed Batch Fermentor

Fed batch process is based on feeding of a growth limiting nutrient substrate to a culture. The fed-batch strategy is typically used in bio-industrial processes to reach a high cell density in the bioreactor. Mostly the feed solution is highly concentrated to

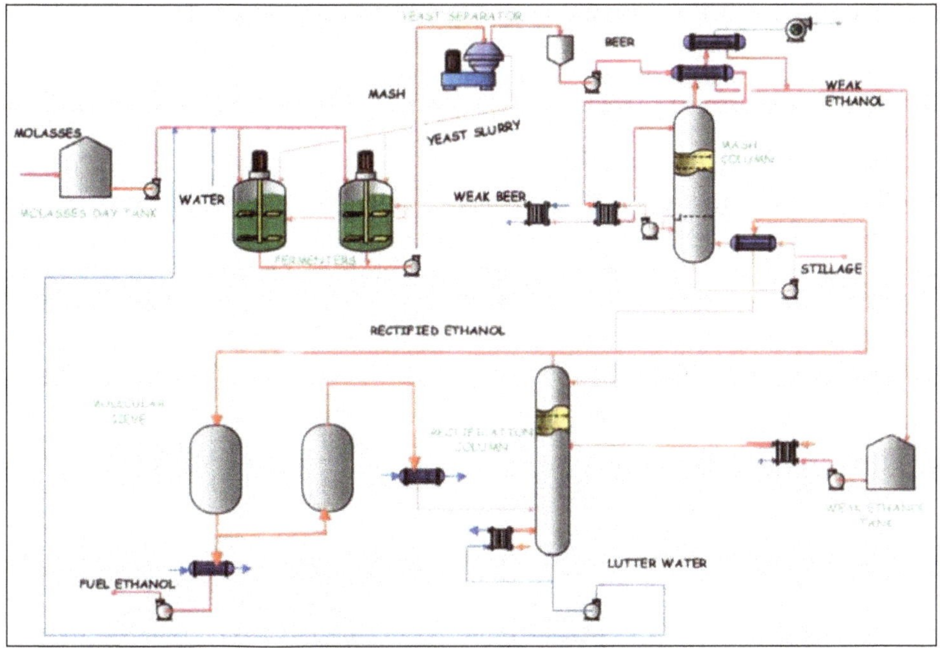

Figure 4.4: Continuous Fermentor

avoid dilution of the bioreactor. The controlled addition of the nutrient directly affects the growth rate of the culture and allows to avoid overflow metabolism (formation of side metabolites, such as acetate for *Escherichia coli*, lactic acid in cell cultures, ethanol in *Saccharomyces cerevisiae*), oxygen limitation (anaerobiosis).

3. Air-Lift or Bubble Fermentor

A fermentor in which circulation of the culture medium and aeration is achieved by injection of air into the lower part of the fermentor. Usually not suitable for animal cell production. Related to gas lift systems where an inert gas is used to achieve circulation in anaerobic conditions. This type of fermentor is well suited for large-scale production of monoclonal antibodies.

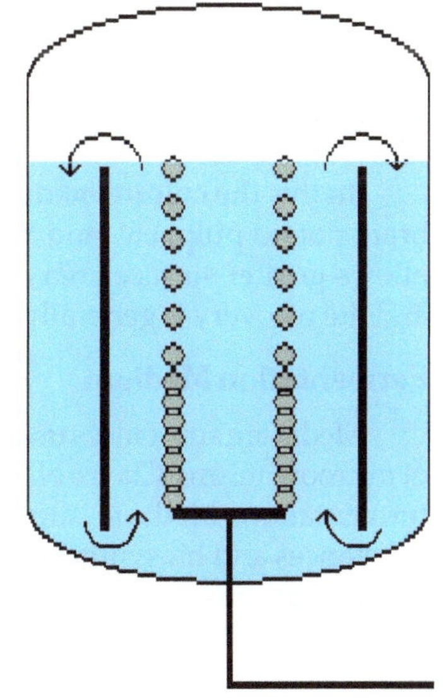

Figure 4.5: Air Lift Fermentor

Types of Culture Method in Fermentation

i. Submerged Culture Method

In this process, the organism is grown in a liquid medium which is vigorously aerated and agitated in large tanks called fermentors. The fermentor could be either an open tank or a closed tank and may be a batch type or a continuous type and are generally made of non-corrosive type of metal or glass lined or of wood. Most fermentation industries today use the submerged process for the production of microbial products.

ii. Surface Culture Method

In this method the organism is allowed to grow on the surface of a liquid medium without agitation. After an appropriate incubation period the culture filtrate is separated from the cell mass and is processed to recover the desirable product. Sometimes the biomass may be reused. Examples of such fermentations are the alcohol production, the beer production and citric acid production. This method is generally time consuming and needs large, area or space.

Figure 4.5: Submerged Culture Method

Figure 4.6: Surface Culture Method

iii. Solid or Solid State Methods

In this the culture medium is impregnated in a carrier such as bagasse, wheat bran, potato pulp, etc. and the organism is allowed to grow on this. This method allows greater surface area for growth. The production of the desirable substance and the recovery is generally easier and satisfactory.

Fermentation Medium

Media are substances used to provide nutrients for the growth and multiplication of microorganisms. Choice of the good medium is important for the success of industrial fermentation. Medium supplies nutrients for growth, energy, building of cells substances and biosynthesis of fermentation products. In laboratory research with microorganisms, pure defined chemicals are used.In industrial fermentation complex, undefined substrates are used for economic reason.

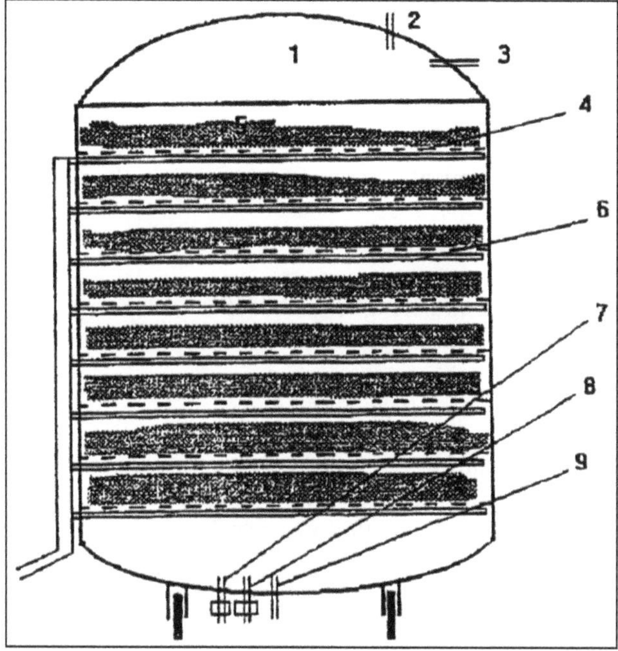

Figure 4.7: Solid or Solid State Method

Media Composition

Among different sources, the particularly important sources are carbon and nitrogen which are the main composition of microbial cells. In addition to carbon and nitrogen media also contains inorganic salts, water, vitamins, other growth factors, buffers, antifoam, lysate of dead cells and fermentation products which are considered as nutrients. In addition to major source and nutrients, media also contains various inhibitors of microbial growth and biosynthesis.

Disadvantages of Poor Fermentation Medium

Poor choice of medium component can cause limited cellular growth, low yield of fermentation product. Poor medium can alter the types and ratio of products from those for which microorganisms has biosynthetic capacity.

Classification of Fermentation Media

Depending on the source utilized fermentation media can be classified as

1. Synthetic media
2. Semi synthetic media
3. Crude media.

1. Synthetic Media

Synthetic medium is a medium in which all of the constituents are specifically designed with known components. Every constituent is relatively pure component.

Exact amounts incorporated into the medium are known. Eg. of synthetic medium composition, inorganic salts, water, purified sugar, ammonium component or nitrate component. Due to the advantages of synthetic medium it is mainly used in research.

2. Semi synthetic media

The media whose chemical composition is partially known as semi synthetic media.

3. Crude media

The crude media allow much higher yield of fermentation products. An example of a crude media is one containing soya bean meal, black strap molasses, corn steep liquor, etc.

Inoculums

Inoculum is the substance/cell culture that is introduced to the medium. The cell then grows in the medium and conducting metabolisms. Inoculum is prepared for the inoculation before the fermentation starts. It needs to be optimized for better performance.

Protocol for Fermentation

Autoclaving

The bioreactor with all accessories (except the stirrer motor) can be paced in an autoclave and it should stay at 121°C for atleast 20 min in order to kill all organisms and thermo resistance spores.

Preparing Cultivation

Reassemble the Bioreactor and liquid addition bottles, after autoclaving.

1. Installation: Put the reactor along side the control equipment (the bioreactor and bioconsole) and connect both pH and DO_2sensors cables to the corresponding electrodes.
2. Preparing for operation: After connecting all cables and tubing, adjust the set points of the pH, temperature and stirrer speed controller to the desired level.
3. Calibrating the DO_2 electrod : After autoclaving the reactor,the DO_2 electrode must be calibrated.

Cultivation

After proper preparation the bioreactor is ready for cultivation. The first step is to create an optimum environment in the reactor by activating the control loops of the bioreactor.

Inoculation

The sterile syringe is filled with inoculum. The needle is pierced through the septum and the inoculum is injected into the rector.

Sampling

To sample a fermentation broth, a sample pipe is needed to which a sample system is attached.

Pasteurization

After completing the fermentation the broth can be pasteurized to kill the organisms. After pasteurization, the heat plate of the reactor can be removed and the fermentation broth can be collected.

Cleaning

After finishing the fermentation process, the glass and stainless steel parts should be cleaned thoroughly using hot water (or) 70 per cent ethanol.

Reassembling

After cleaning, dry the parts and reassemble the reactor.

Functions

1. Provides operation free from contamination.
2. Maintain a specific temperature
3. Provide adequate mixing and aeration
4. Control the pH of the medium
5. Allow monitoring and/or control of DO
6. Allow feeding of nutrients, solutions and regents
7. Provide access points for inoculations and sampling
8. Use fittings and geometry relevant to scale up.
9. Minimize liquid loss from the vessel.
10. Facilitate the growth of a wide rage of organisms.

Application of Fermentors in different Fields

1. *Useful Intermediate Products*: The various intermediate compounds produced during fermentation activity are classified as primary and secondary metabolites. Some of the commercially important metabolites are (a) Primary metabolites, such as amino acids, proteins, carbohydrates, vitamins, acetone, ethanol, organic acids, etc. and (b) Secondary metabolites such as antibiotics, toxins, alkaloids, gibberellins, etc.
2. *Enzymes*: Enzymes produced by the micro-organisms during fermentation include amylases, cellulases, invertase, esterase, lipase, protease, lactase, etc.
3. *Microbial Biomass*: After the process of industrial fermentation is over, the exhausted cells of the micro-organisms (in the fermentor) serve as microbial biomass. It is also known as 'microbial protein' or 'single cell protein' (SCP). It is an important source of proteins.

4. These devices are being developed for use in tissue engineering.
5. Biological hydrogen production is done in a bioreactor based on the production of hydrogen by algae.
6. Involved in the cultivation of microbes for the production of SCP.
7. It plays a role in exploiting the metabolites such as ethanol, acetone, butanol, lysine, glulamic acid and polysaccharides from the microbes.
8. It allows microbes to grow for the production of enzymes.
9. Algae can be cultured in mass or continuously.
10. Involved in the production of recombinant protein and vaccines.\
11. Large scale cultivation of mammalian cells can be achieved through the bioreactor.
12. It enhances the nutritional value of the feed ingredients.
13. Used to produce Marine single cell detritus (MSCD) an alternative diet for shrimp larvae in hatcheries.

References

Caplice, E. and Fitzgerald, G. F. 1999. Food fermentations: role of microorganisms in food production and preservation. *Int. J. Food Microbiol.* 50:131–149.

Fleming, H. P., Kyung, K.H. and Breidt Jr, F. 1995. Vegetable fermentations, *In* H.-J. Rehm and G. Reed (ed.), *Biotechnology.* VCH Publishers, New York. pp: 629–661.

Hunter, I. S. 1999. 'Fermentation Microbiology and Biotechnology' ed. E. M. T. El-Mansi and C. F.A. Bryce, Taylor and Francis, London. pp: 121.

Takeda, H. *et al.,* 2011. Bioethanol production from marine biomass alginate by metabolically engineered bacteria. *Energy Environ. Sci.* 4: 2575-2581.

Uchida, M. 2002. Fermentation of seaweeds. *J. Jap. Lactic Acid Bac.* 13: 92-113 [In Japanese with English summary].

Uchida, M. and Murata, M. 2004. Isolation of a lactic acid bacterium and yeast consortium from a fermented material of *Ulva* spp. (Chlorophyta). *Journal of Applied Microbiology.* 97: 1297-1310.

Uchida, M. *et al.,* 2004. Mass preparation of marine silage from *Undaria pinnatifida* and its dietary effect for young pearl oyster. *Fish. Sci.* 70: 456-462.

Chapter 5

Marine Single Cell Detritus: Principle and Protocol

S. Felix

Fisheries Research and Extension Centre,
Tamil Nadu Fisheries University,
Chennai – 600 051, T.N.

Innovative aquaculture products for bioremediation for attaining biosecurity are increasingly being produced all over the world today. Nutrition requirement for marine larval development is one tricky area evading solution to many problems during the past few decades. Attempts are on to replace phyto and zoo plankton cultures that form the single source of nutrition to sustain the marine larval production till date. The algal concentrate however could not be depend upon as their production is highly unreliable. Many technologies have been developed so far to replace algae, which include microencapsulated diet to replace algae in marine hatchery technology. The marine single cell detritus (MSCD) production is presently gaining considerable attention of aquaculture scientists to replace or substitute algae in marine hatcheries.

Single Cell Detritus (SCD) and their Role

Fermentation is one of the oldest biotechnological aspect which could be exploited for larval feed (MSCD) preparation. Motoharue Uchida of Japan was the initiator of the formulation and production of MSCD for marine hatcheries. For maximum utilization of the dietary potential of micro algae, it is advantageous to perform thalli degradation under conditions regulating the catabolic losses. Mechanical or enzymatic fragmentation is effective for this purpose. Conversely, a method using viable bacteria for degradation is advantageous. Another interesting characteristic of the detritus diet is the attachment of bacteria to the surface of the detritus. It is also possible to supply the Single Cell Detritus (SCD) particles with different bacteria attached to its surface by incubating the bacterium for several hours axenically prepared SCD particles. This optional method of attaching bacterium has some useful functions such as anti-pathogenic activity, or a vitamin - producing ability, is expected to be useful in the development of a functional hatchery diet for suspension feeder animals. The combination of lactic acid bacteria and yeast might have a synergetic effect for reducing the risk of the prevalence of pathogenic microbes in the production process. Furthermore, some studies also suggest the possible probiotic effect of lactic acid bacteria and yeast on aquatic organisms.

What is Marine Single Cell Detritus (MSCD)

MSCD is the seaweed based product which is produced through a combination of enzymatic and fermentative techniques. It can be prepared at the particle size of 5 to 12µm, making them ideal for marine hatchery feeding apart from its bioremediatory and probiotic roles.

Production of MSCD

MSCD production has two phases. The first phase is cellulolytic enzymatic treatment of seaweed which leads to single cell units. The enzymatic digest is further treated with bacteria and yeast in the second phases can be performed simultaneously or one by one.

1. Cellulolytic Enzymatic Phase

The microalgae have cellulose in their cell wall which keep the cells intact. When cellulose is digested, the individual cells are released and become single cell units. The enzyme 'cellulase' is used for the purpose and the end product of cellulolytic digestion is sugar. This phase thus has two roles, one is to produce single cell units and other is to produce sugars for the fermentative phase.

2. Fermentative Phase

Two organisms are used in this phase to produce MSCD and they are lactic acid bacteria (LAB) an yeast. These organisms can be isolated from the natural fermented seaweed or from other sources. Motoharue Uchita (2004), used *Lactobacillus brevis* (LAB), *Debaryomyces hanseii* var *hanseii* (Yeast) and *Condida zeylanoides* (yeast) that are isolated from fermented *Ulva*. Bacteria like *L.plantarum* and *L.casei* can be used for this purpose. In case of yeast, any suitable source of yeast can be used for fermenting the seaweed.

Lactic acid bacteria and yeast utilize the sugar produced by cellulolytic digestion and produce lactic acid and this would prevent other organisms from growing and thus preserving the MSCD. MSCD can be sieved with appropriate filters to obtain the required size for feeding. It can be stored for a year at room temperature. LAB also acts as a probiotic and thus help to increase the survival and to maintain the water quality. Yeast predominately acts as a bioremediatory agent and enable us to maintain the culture systems as a zero water or limited water exchange system.

Figure 5.1: ***In-situ*** **Fermentor (10 L)**

Figure 5.2: MSCD Preparation

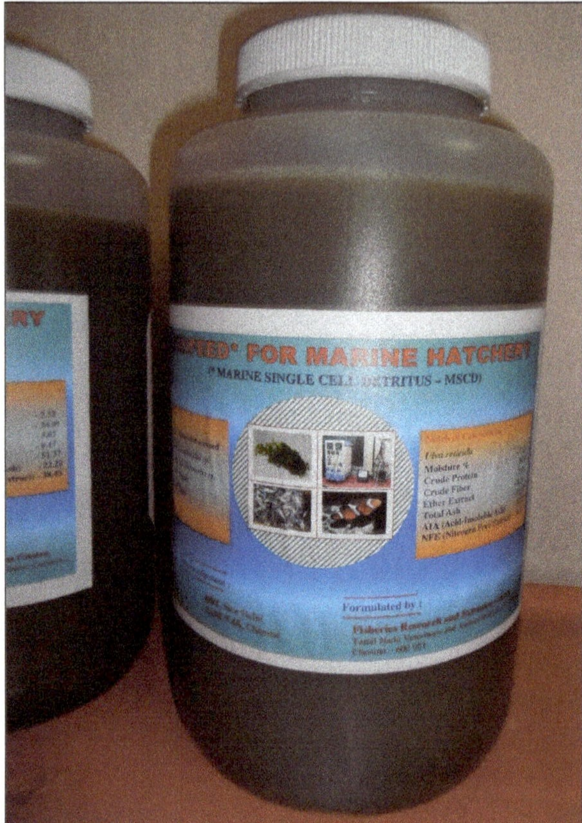

Figure 4.3: MSCD Product

Large Scale Production

MSCD can be produced in large quantities in air tight containers or sophistically in 'Fermentor' or 'Bioreactor'. The difference between the two is that in air tight containers it takes nearly 2 weeks to ferment but it takes only a maximum of 3 days to ferment in a fermentor. Further, for purity and for production of quality MSCD, fermentors are recommended.

Application of MSCD

MSCD can be fed to larvae of both finfish and shellfishes. The application of MSCD as feed has been tried in oyster and *Artemia* by Motoharue Uchita in Japan. Presently our lab is working towards the formulation and production of MSCD for shrimp larvae (*Penaeus monodon*), as a replacement for unicellular algae and our trials so far have been successful. The expected break through would make a turn around in shrimp hatchery feeding technology. The shrimp hatchery nutrition management would become simple and cost-effective.

Advantages of MSCD

- ☆ Relatively nutritious (crude protein levels upto 30 per cent has been reported).
- ☆ Effectively could replace (partially or fully) the microalgae as a feed in hatcheries.
- ☆ Particles of required size can be produced, as per our need and species which we are interested to produce.
- ☆ Can be stored even for a year at room temperature.
- ☆ They act as a bioremediatory agent
- ☆ The probiotic effect also has been established.
- ☆ The high cell concentration of MSCD is comparable with algal concentrates.
- ☆ Better utilization of seaweed resources through MSCD production.
- ☆ Mass preparation of MSCD is relatively easier than the production of microalgal culture and its maintenance.
- ☆ Economically viable technology.

Marine Single Cell Detritus (MSCD) technology is here to change the way aquatic organisms are fed, particularly in marine hatcheries in the near future. The technology integrates nutrition, probiotic and bioremediatory aspects together for the better nutrition management in hatcheries. Hence it is an innovative, multifaceted and integrated technology throw more light on the seaweed resource utilization and also would paveway for simplifying the marine larval rearing technology.

References

Felix, S., T.V.Shivakumar, Samayakannan M. 2005. A prototype nursery raceway and low water exchange growout system for shrimp. *INFOFISH International.*, 5: 14-19.

Felix. S, R. Ramya and Shanmugam S.A. 2007. A MSCD substitute diet for microalgae in marine hatcheries. *INFOFISH International*, 3 : 8-11.

Felix S., P. Pradeepa and Shanmugam S.A. 2011. Relative digestion efficiency of cellulase enzymes on seaweeds of Gulf of Mannar for their utilization in Marine Single Cell Detritus (MSCD). *Seaweed Res. and Utiln.*, 33 : (1&2) 77-82.

Fleurence J. 1999. The enzymatic degradation of algal cell walls: a useful approach for improving protein accessibility. *J. Appli. Phycol.*, **11:** 313–314.

Hall, G.M. and Silva, D. S. 1992. Lactic acid fermentation of shrimp (*Peneaus monodon*) waste for chitin recovery. *In*: Advances in chitin and chitosan. Brine C J. Sanford P S. Zikakis J P.(eds). *Elsevier Applied Sciences*, London.; 633 – 668.

Nikolaeva E.V., A.I. Usov, A.P. Sinitsyn and Tambiev A.H. 1999. Degradation of agarophytic red algal cell wall components by new crude enzyme preparations. *J. Appli. Phycol.*, 11: 385–389.

Ramya R. and Felix S. 2009. Preparation of Marine Single Cell Detritus (MSCD) using automated fermentor for nursery rearing of shrimp. *Seaweed Res Utiln.*, 31 (1&2) : 227-234.

Siddhanta A. K., A.M. Goswami, B. K. Ramavat, K.H. Mody and Mairh O.P. 2001. Water soluble polysaccharides of marine algal species of *Ulva* (Ulvales, Chlorophyta) of Indian waters. *Indian Journal of Marine Sciences.* 30: 166-172.

Uchida M., 2003. Use of fermented seaweed as a hatchery diet. *Aqua Feed International*, 2: 15-17.

Uchida M. 2005. Studies on lactic acid fermentation of seaweed. *Bull. Fish. Res. Agency*, 14: 21-85.

Uchida M. and Murata M. 2002. Fermentative preparation of single cell detritus from seaweed, *Undaria pinnatifida*, suitable as a replacement hatchery diet for unicellular algae. *Aquaculture*, 207: 345-357.

Uchida M., K. Numaguchi and M.Murata. 2004. Mass preparation of marine silage from *Undaria pinnatifida* and its dietary effect for young pearl oysters. *Fish. Sci.*, 70:457-463.

Uchida M and Murata M. 2004. Isolation of a lactic acid bacterium and yeast consortium from a fermented material of *Ulva* spp. (Chlorophyta). *J. Appli. Microbiol.*, 97: 1297-1310.

Part III

Bioremediation and Biosecurity in Aquaculture

Chapter 6

Biosecured Raceway Technology

S. Felix

Fisheries Research and Extension Centre,
Tamil Nadu Fisheries University,
Chennai – 600 051, T.N.

Raceway technology was standardised for the first time in India for the shrimp rearing under the DBT funded project at Fisheries College and Research Institute, Thoothukudi. The feasibility of utilizing this system for ornament fish rearing also has been proved successful here at Fisheries Research and Extension Centre, Madhavaram. This chapter deals with the raceway technology as standardised for shrimp. The same technology however can be domesticated for ornamental fishes with suitable modifications.

One of the major problems facing the shrimp farming industry is poor survival and production predictability hen juvenile shrimp (PL) are stocked into grow out ponds. To tackle this, an advanced management strategy has been developed to increase juvenile shrimp survival and production predictability for 50 days through raceway rearing of post larvae and then growing them in Limited Water Exchange Grow out (LWEG) ponds for further culture period. The raceway technology to raise shrimp crop has been put to use effectively and successfully for past few years in countries such as USA, Mexico, Hawaii and in few South East Asian Countries (Samocha, *et al.*, 2000 and 2002). It has been proved that raceways with limited/zero water exchange can be used to raise shrimps as a full crop or in a two phase systems, in order to help to minimize the crop loss (Felix, S. 2001). Nursery systems have advantages over direct stocking, which increases control over stock inventories, water quality and feed management (Samocha and Benner, 2001). It provides for larger and hardier shrimp to be stocked and facilitates a higher number of crops per pond as a shorter period is needed to reach a marketable size (Clifford, 1985; Briggs and Brown, 1991; Fast, 1991). The survival of this nursery system is high (85 – 95 per cent) with higher anticipated profit (Hirono, 1983; Aquacop, 1985; Fast, 1991; Sturmer *et al.*, 1992, Samocha and Lawrence, 1992). Use of nursery systems can also prevent the spread of diseases as post larvae can be kept under quarantine during the nursery. Thus, a significant improvement in yields has been reported from white spot infected areas in Ecuador when intensive nursery facilities have been used rather than direct stocking (Samocha *et al.*, 2000 a, b; Samocha *et al.*, 2001). Farms with intensive nursery facilities can also stock and store PLs in high density to reduce pressure on limited hatchery supplies (Samocha and Benner, 2001). The University of Texas at Austin, Marine Science Institute, Fisheries and Mariculture Laboratory, has employed water reuse systems for over 25 years for both research and intensive production of fish and shrimp. Arnold *et al.* (1990) and Reid and Arnold (1992) provide a general description of the basic recirculating system used for shrimp production. Davis and Arnold (1998) have reported about the nursery raceway system with detailed description of this nursery raceway system along with several alterations of procedures and system design to increase the efficiency of waste removal, to minimize labour requirements and to maximize biomass loading. Otoshi *et al.* (2002) used the recirculating raceway systems for the production of brood stock shrimps (*Litopenaeus vannamei*).

The techniques described here are prototyped and demonstrated at the Fisheries College and Research Institute, Thoothukudi for the first time in India. An advanced management strategy has been developed to increase juvenile shrimp survival and production for the initial 50 days (the crucial phase of shrimp culture) in tropical Indian conditions. The operation procedures described herein have resulted in high survival (above 80 per cent in 50 days) at a stocking density of 2000/m^3. Nevertheless, the potential carrying capacity of this system is even higher, more than 6000/m^3 (Qirong, W., 2003 and Samocha *et al.*, 1993). Today, in India there are no nursery systems adopted for shrimp rearing. This write up describes design and construction of raceways, operation and management of raceways and economic feasibility of this system. It would throw light and guide the farmers/technologists to adopt intensive nursery raceway systems effectively for shrimp rearing, which has been proved to be eco-friendly and economically viable.

Design and Construction of Nursery Raceway System

Major components of a raceway system:

1. Raceway tanks
2. Greenhouse structure for raceways
3. Limited water exchange grow out systems
4. Storage and mixing tanks
5. Filtration systems (Biological filter, U.V. and Rapid pressure sand filters)
6. Generator, Blowers and Paddle wheel aerators
7. Indoor and outdoor algal culture facilities

8. Farm Laboratory
9. Sea or saline water and Pumping facility

1. Raceway System

To construct a raceway of 15 X 3 x 1 m dimension, required earth was excavated and the dikes were formed for the raceways and after sufficient consolidation, fabricated pre-cast cement slabs (50mm thickness) were fixed on the bottom and the sides of the raceway pond. These elongated raceways (15m long; 3m width; 1 m depth) with rounded corners were lined with 700 GSM nylon fabric sheet. Water outlet drain provision was given to tanks towards the end of raceways. Bottom of the raceway tanks were provided with an adequate slope (0.5 per cent) towards the drain to allow easy draining of water. The outlet of raceway tanks leads to harvesting tank (3 x 1.5 x 2.1 m) and the used water was drained to constructed wetlands (CWs). The depth of the raceway was kept as 100 cm. Freeboard of 15 cm was provided for producing a maximum of 3-5 g size shrimps in raceway tanks.

Each raceway was provided with a 13 m long and 1 m deep central partition (baffle). Central partition was made up of waterproof marine plywood of 20mm thickness and fixed on cross-wooden beams with stainless steel bolts and screws. The height and position adjustment of central partition was also provided.

2. Greenhouse Structure for Raceways

A greenhouse was constructed using cast iron pipe pillars, framing with gentle semicircle roofing of Iron pipes coated with non-corrosive paint. The dimension of

Figure 6.1: Fixing of Central Partition

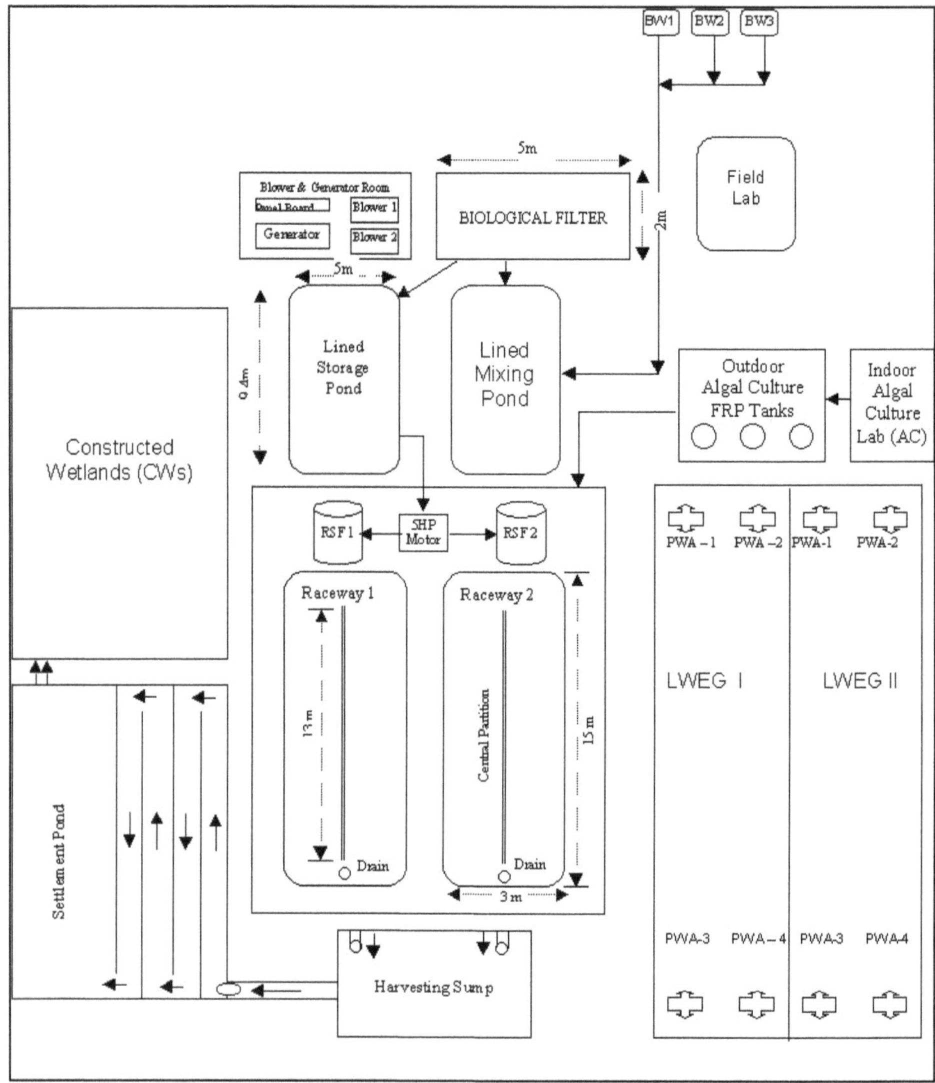

Figure 6.2: Design Layout of Raceway Farm Complex (Prototype)
LWEG: Limited Water Exchange Growout Pond; RSF: Rapid Sand Filter; PWA: Paddle Wheel Aerators; BW: Borewell

greenhouse was 20 x 10 m. The roofing was provided with 20 per cent of transparent FRP sheets and 80 per cent of non-transparent white FRP sheets. The sidewalls were made up of removable green shading (75 per cent) net material fixed to wooden frames. The structure is large enough to maintain two 30m^3 raceways with pumps and two sand filters. The greenhouse space is equipped with 20 tube light fittings arranged in four rows. Building raceways facilities under greenhouses would help in maintaining the ambient water temperature and other water quality parameters like DO, salinity, pH, ammonia, nitrite, microbial load, algal density, etc.

Figure 6.3: Roofing of the Greenhouse with FRP Sheets

Raceway Outlets and Filter Pipes

The drain outlet of raceways was located towards the drain end, half way between the end of the partition and the raceway's end wall. The raceway water level can be controlled by an external standpipe positioned inside the harvesting tank. Each raceway should be provided with a different set of perforated filter pipes covered with following screen sizes 600, 800, 1000 and 2000 µm. Perforated filter pipes are mounted on the outlet to avoid losing shrimp during water drain. Filter pipes can be changed as the shrimp grow.

Figure 6.4: Outside View of the Raceway System

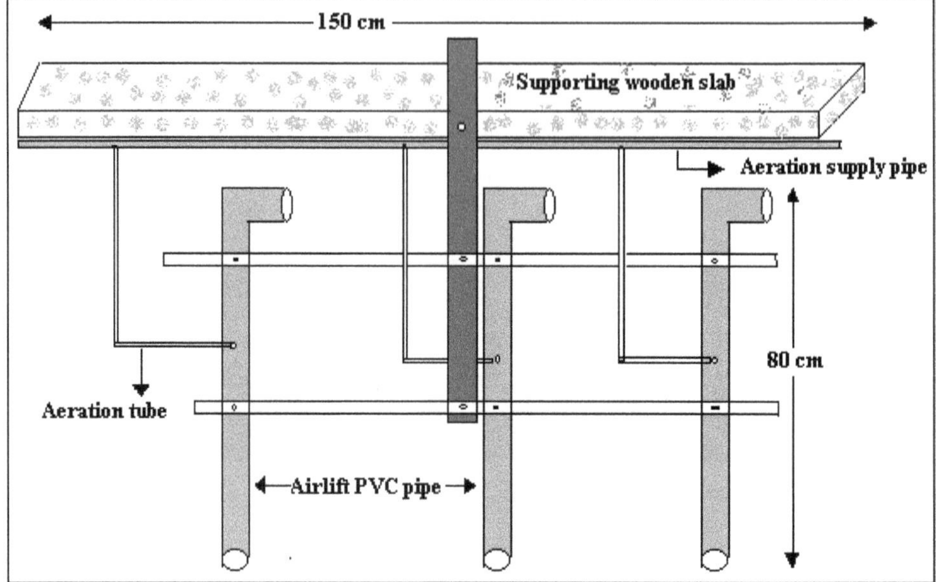

Figure 6.5: Airlift and Structural Support (LS).

Harvesting Tank

Raceways should be designed in such a way to allow drain harvest. Towards the drain end of two raceways single harvesting tank (2x1.5 x 2) was provided for the purpose of water exchange, bottom waste clearing and for harvesting. Outlet drain pipes of both raceways were connected to harvesting tank, and outlet of harvesting tank in turn was connected to the constructed wetlands (CWs). During water exchange for waste removal and harvesting 1000 to 2000 μm nylon net screen frame were used inside the harvesting tank (as per the size of shrimps) to avoid escape of shrimps.

Air Lift System

Water circulation in the raceways is depending primarily on airlift systems. In this system, air is introduced 30 cm below the water level inorder to attain a maximum uplift of water through a vertical PVC pipe which has a 90° PVC elbow at the upper section. The air will lift the water through the pipe from bottom and sends it through the elbow in desired direction by which unidirectional water flow was ensured. Air supply is ensured by three 5 HP blowers that operated alternatively every 3 hours. Air pressure, airlift pipe diameter, submergence depth and the type of airlift system being used are the major factors affecting the pumping rate and water circulated. In the raceway, airlift systems are arranged in 6 sets, 3 on each side of a central baffle and each set was provided with 3 airlift pipes. The airlift pumps are fixed to cross-wooden beams with adjustable stainless steel screws, which will allow in adjusting the height of the airlift pumps in raceways when water level is low.

Rapid Sand Filterss

A rapid sand filter (Waterco, Australia) with manual backwash and filtration capacity of about 20000 LPH was provided for each raceway. The sand filter can be used to filter the incoming seawater as well as the raceway water. The multiport valve in the rapid sand filter is a multi-position valve with six operational modes *viz.* sand filter, back wash, rinse, circulation, waste and closed. The backwash and rinse modes serve to maintain the sand filter in optimal working conditions. The circulation bypasses the sand filter. The waste mode can be used to drain the raceway without using the external standpipe. The closed mode is a safety position to avoid accidental drainage through the multi port valve when the raceway is not in operation.

Figure 6.6: Rapid Sand Filters Installed in Raceways

UV Sterilizing System

Two numbers of UV sterilizers (Rainbow Lifeguard, India) were installed to treat the incoming water to the raceways. UV sterilizers were connected in the water intake system and the filtered water from the rapid sand filters was passed through these UV sterilizers.

Biological Filter

A conventional gravity flow biological filter was provided for the intake water. Biological filter tank, having the capacity of 5 x 2 m with central partition and three outlets (50mm) positioned below the bottom of central partition. In this one part was provided with different layers of filtration material above the false bottom provided *viz.* big and medium size gravels, smaller size gravels, activated charcoal, coral sand,

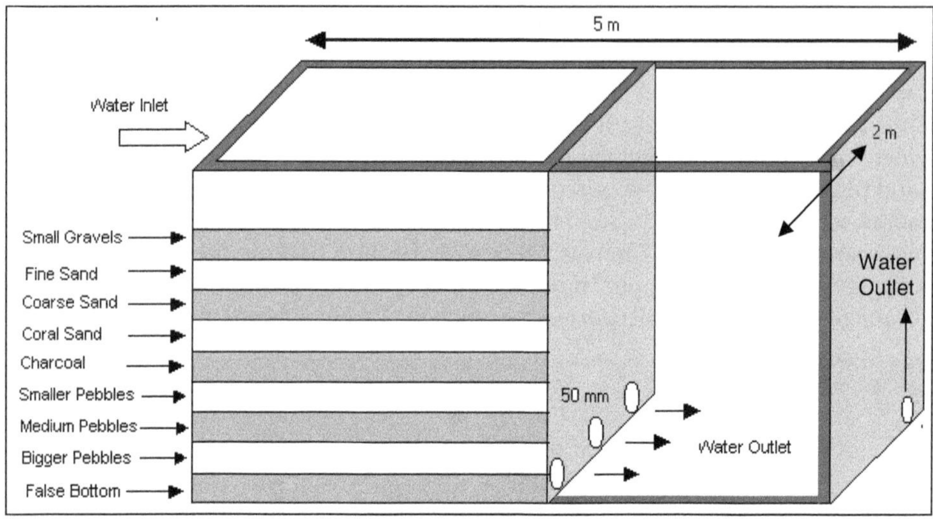

Figure 6.7: Lagers of Biological Filter

coarse sand and fine sand. A nylon net material separated each layer. Water will flow through the filtration tank to the other side partition that is into the empty tank and from there to LDPE lined storage pond.

Storage and Mixing Ponds

These ponds were constructed with a dimension of 10 x 5 m each. The dikes and bottom were lined with 250 GSM LDPE sheets to arrest water seepage. Water from all the three borewells (different salinities) is pumped to the mixing pond, where the required salinity is adjusted and from there it is pumped to the biological filter. The biofiltered water is then stored in the storage pond. From there water is pumped to raceways by a 5HP motor through U.V. and pressure sand filters (20,000 LPH).

Acclimation Tank

A few days before the scheduled post larvae (PL) arrival, acclimation tanks should be cleaned and disinfected. On the day of stocking, tanks should be filled with filtered seawater. Water temperature and salinity have to be adjusted according to the anticipated salinity and temperature in which the post larvae are transported.

Algal Culture Facility

Indoor Algal Culture Facility

An indoor algal culture facility was established with air-condition facility to maintain optimum temperature. Walnae's (f/2) medium was used to culture *Cheatoceros calcitrans* diatom. At regular intervals transfer of inoculum from lower dilution to higher dilution was carried out to sustain the algal production in the indoor culture. For culturing indoor algae 50ml, 100ml, 250ml, 500ml and 1000ml conical flasks with 4 litre LDPE pet jars and 20 litre capacity transparent buckets were used. Before addition of inoculum, every stage of algal growth was carefully observed under microscope. In indoor algal culture, sterilization of seawater and glasswares were done by autoclaving or by heat sterilization of water using heaters. Water needs to be cooled to optimum temperature before the addition of materials and inoculum. Required aeration and light illumination need to be provided round the clock through air pumps and 40W tube-lights respectively.

Figure 6.8: Application of Algal Concentrate

Outdoor Algal Culture Facility

Outdoor algal culture was carried out in 500 - 1000 Litres transparent FRP tanks. Indoor produced *Chaetoceros calcitrans* diatom is inoculated to outdoor FRP tanks and filtered water was fertilized with f/2 Walnae's medium. After 24-48 hours depending on cell density, algae from FRP tanks were transferred to raceway tanks or LWEG ponds by pumping (0.5 HP). Cell density and quality were estimated each time under microscope.

Blowers and Generator

Application of algal concentrate three numbers of twin lobe air blowers (5HP each) are used in raceways (two are used alternatively and one is kept as stand-by) for the purpose of aeration and operating airlift systems, which will enhance the dissolved oxygen level in raceways and also enables continuous water circulation.

A 15 KVA generator (KIRLOSKAR) is installed to ensure continuous operation of blowers, paddle wheel aerators and pressure sand filters during power failure.

Alarm System

Simulation of power failure could immediately activate the alarm system. The alarm set is connected to the field lab, which is at the centre of the farm site. When operating personals/technicians get the power failure alarm signals, they can switch on the generator.

Field Laboratory

Routine water quality parameters like pH, Dissolved Oxygen, ammonia and nitrate, etc. were tested in Field laboratory and also the routine health check up of shrimps, feed rationing, data register maintenance etc. are under taken in the field lab.

Shrimp Culture Raceways: Concept, Working Principle and Relative Advantages

Global shrimp culture industry is now facing a major challenge of its sustainability due to dreadful diseases and deteriorating environment. The effective strategy to tackle the problem is biosecurity that is to physically isolate the farm from its surroundings and preventing waste accumulation through bioremediation and limited or zero water exchange. It is an economically viable system or method for intensive production of high quality disease free shrimps, which also minimizes environmental impacts. Biosecurity should have several layers of protection and planning to accommodate the possibility of stressors and disease penetration. It is envisaged that working with biotechnology fields will bring together emerging aquaculture needs and possibilities with new approaches, methodologies, sensors and advanced understanding of such complex systems.

The biosecure zero-exchange system described here is currently leading a technology revolution in aquaculture worldwide.

The raceways are differentiated as follows :

1. Traditional raceways are enclosed channel systems with relatively higher rates of moving or flowing water. Water flowed through the system by gravity, but not pumped.
2. Pond raceways are the viable ones, in water shed ponds where raceways are constructed out of plywood and plastic sheets and provided with an air lift system, waste reduction system and with high water exchange rates.
3. Greenhouse raceways are managed with zero to minimal water exchange, thus greatly reducing environmental impacts showing high survival and higher production.

Figure 6.9: Raceway in Operation (Texas, USA).

4. Inland raceways are maintained in inland areas with addition of fresh water and brine and adopting greenhouse system and zero water exchange.
5. Biosecured raceways involves a combination of physical, chemical and biological measures like use of specific pathogen free shrimp stocks, treatment of incoming water, greenhouse system, zero water exchange, filters, air lift system, bioremediation in culture system, etc., to ensure high survival and production.

Concept

Shrimp farming technology is currently taking up an entirely new dimension with biosecured nurseries, super intensive greenhouses, zero water exchange and integration with constructed wetlands. More specifically, the system comprises a synergistic interaction between algae and bacterial population in an aqueous medium for in-situ waste treatment and the production of a live nutrition source, shrimp free from pathogens, and an adaptable nutrition source to support water quality and prevents the buildup of toxins.

1. For better survival and yield of shrimp larval rearing, raceways are built from concrete, fibreglass, plywood or simply by lining with HDPE membrane under a greenhouse system.
2. Zero water exchange system to enhance shrimp production.
3. Bioremediation using microbial, algal and fermented products supplements as a flocculent.

The system has been divided into its component parts, namely: zero water exchange system, an aqueous medium, the microbial population, the shrimp population, a growout tank and associated equipment, a nutrition source, and a method for using the system to achieve shrimp growout.

Zero Water Exchange System

The biosecure system accomplishes the intensive culture of shrimps with zero aqueous medium exchange ("zero-exchange"). The concept of zero-exchange refers to a system wherein new aqueous medium is introduced into the system only to replenish water lost for physical reasons, specifically evaporation and experimental sampling and not for chemical or biological dilution. The concept of zero-exchange also relates to metabolites and solids provided to, or formed in the system during growout. If these solids and metabolites are retained in the system during the growout cycle, it would be critical for the growth of the shrimp stock and for maintenance of a synergistic microbial population. The systems recognize the importance of maintaining solid residues, faecal matter, particulate matter, metabolites and uneaten feed in an intensive zero-exchange growout system to provide an environment favorable for high growth rates and yields of shrimp. The system is sufficiently isolated to prevent the introduction of pathogens to the microbial population and in the presence of a zero-exchange during the entire growout cycle. When combined, these individual components form a balanced system, wherein the shrimp growth rate will be increased by 100 per cent to 500 per cent, relative to "clean water" systems, and provides low cost production of high quality, disease free and commercially desirable shrimps.

Bioremediation in Raceways

Bioremediation is a biotechnological process of using selected micro or macro organisms to reduce harmful wastes to less hazardous levels. The principle is that to use selected microorganisms outside the host to create the healthy environment. *Bacillus* sp, *Vibrio* sp, *Nitrosomonas* sp, *Nitrobacter* sp, *Rhodopseudomonas* sp, *Aerobacter aerogenosa, Cellulomonas biazotea, Saccharomyces* sp, several algae and ferns are biological agents recognized for their bioremediation abilities.

Accumulation of unconsumed high protein feed, shrimp excretion and microbial degradation processes of organic matter result in increase of ammonia level. Introducing active aerobic bacterial populations could be an effective strategy in rapid degradation of complex organic compounds and overall improvement of health status of cultured organisms.

Algal inoculation as well as introduction of indigenous fermented products also will be of more use.

The Aqueous Medium

The building block of the aqueous growth medium is specific pathogen free water. The aqueous medium is based upon water that is disinfected, and further treated to provide a medium suitable for the growth of microorganisms and shrimp. The aqueous medium is usually seawater, which has a suitable level of salinity. By reducing the salinity of the aqueous medium to levels below that of seawater and by eliminating the contaminants in the system, the requirement to use ocean water to grow shrimp is eliminated. This enables the production of shrimp farming facilities inland, away from coastal zones. Therefore, the water used in the aqueous medium may come from a variety of sources. For example, the water may be open well water, subsoil saline water from bore wells, river or lake water, spring water, brackish water, or even tap water. Additionally, the acceptability of water salinities below sea water further simplifies water disinfection by enabling the use of disinfected fresh water (tap water) with disinfected sea salts, or a synthetic imitation thereof, to provide a suitable level of salinity and mineral content.

The Raceway Tank and Associated Equipment/Structure

The shape/configuration of the raceway tank in which the aqueous growth medium and shrimp are deposited is largely a matter of choice. Important factors to consider in tank design for the present system are the cover, isolation of the tank, oxygenation, availability of a light source and circulation. If covered or housed in a building, however, the tank cover or building ceiling should be wholly or partially transparent in order for light, including photosynthetically active light, to pass through at levels sufficient to insure the health of the microbial and shrimp populations in the growout tank throughout the growth cycle. Alternatively, artificial light, approximating natural light in spectrum, intensity, and cycle duration, may be provided if natural light cannot be made available to the growout medium.

Further, the depth of the tank also affects the growth of the algae population. The algae are important for the growth of the shrimp. If the tank depth is too shallow, light energy will be wasted. If the tank is too deep, the lower portion of the microbial population will not receive the necessary amount of light, thus shifting the system undesirably towards a predominantly bacterial population. The determination of the correct depth requires light attenuation studies of the specific system conditions, tank configuration, and water column depth, and may be accomplished by using a light meter or similar measuring device.

The bottom of the tank and the walls of the tank must be lined to exclude possible disease vectors; to prevent leakage, to facilitate cleaning or disinfection. The aqueous medium should be aerated and agitated. Typically, this is accomplished by circulation of the aqueous medium with a sufficient volumetric flow rate to keep solids formed in or introduced into the growth medium in suspension. Such flow generating devices such as air lift system may contribute to or provide sufficient turbulence to provide adequate aeration of the aqueous medium.

Environmental Impact

Aquaculture wastes are a source of nutrient pollution and can stimulate blooms of toxic algae, which can contaminate shrimps, and potentially affect the production and survival. Further, the introduction of non-native and/or selectively bred animals with wild populations can harm local ecosystems by altering species composition or reducing biodiversity. Additionally, the transmission and spread of disease, is of great environmental concern. Biosecure zero-exchange systems do not contribute to the deterioration of coastal or estuarine ecosystems and their use will also help to reduce pressures on the shrimp capture fisheries and to alleviate the destructive bycatch (*e.g.*, turtles) problems of the shrimp trawler industry.

Advantages of Raceway Culture System

1. Increased yield from raceways depending upon its usage (as nursery or grow out or both) is possible.
2. It works like an indoor shrimp production unit, thus avoiding many problems associated with outdoor shrimp farming systems.
3. The relative quantity of water used is very meager and thus pollution related problems such as salination to adjacent land and water sources is nil.
4. The zero or limited water exchange concept adopted in raceways helped in raising shrimp in an eco-friendly system.
5. Less land and less water means the cost of production is considerably reduced.
6. The system is fully in ones control, implementing bioremediation concepts in raceways will be easier and effective.
7. The 'biosecurity' concept can be adopted more effectively in raceway system.
8. High yield and high survival means the raceway system to produce shrimp is a proven viable technology.

9. The high yield ensures that the system works with in the 'economically feasible' range.
10. Raceway ensures 'disease-free shrimp crops' which is essential for such intensive culture system.

Pre-Stocking Management of Raceways

Raceway Cleaning, Disinfection and Maintenance

Cleaning and disinfection are the important factors for successful nursery rearing operation. After every harvest/at the end of each harvest, all raceway components should be thoroughly cleaned. All fouling organisms and algal biofilms should be removed by using brushes. Calcareous/salt deposits could be removed by applying diluted Hydrochloric acid (Hcl). However human health care should be taken to avoid any hazard. After washing, raceways should be disinfected with a chlorine solution (30 – 100 ppm active chlorine).

Airlift systems were removed from the wooden beams and should be dipped in concentrated Hcl for removing fouling organisms that adhered inside the airlift pipes. Rusted clamps and bolts were removed and replaced with new one after each crop.

The better maintenance of rapid sand filters require operating the sand filter in backwash mode before running each and every time. However, if the organic load is high, thorough cleaning will be needed after each crop; otherwise, servicing of rapid sand filters will be carried out for thorough cleaning. The entire filter media should be removed, washed with purified water and screened to separate filter media based on its particle size. Loss of filter media (sand) can be replaced and repacked.

Raceway Filling and Chlorination

Raceways should be filled with filtered seawater upto the required mutes level (0.15 m) one week prior to stocking. Water was drawn from three different subsoil borewells (as access to sea water is not possible at Fisheries College and Research Institute – the project site) having different salinity and pumped to the mixing pond to adjust the required salinity. The water is passed through a biological filter, rapid sand filter and activated carbon filters and U.V.sterilizing system. Culture water should also be treated with liquid chlorine (10 ppm active chlorine). After 5 – 6Hr, water is vigorously aerated until all chlorine gets liberated. Periodic water sample should be taken and checked for chlorine level. No chlorine level should be detected before stocking/fertilizing the raceways.

Raceway Stocking Management

The quality seeds are to be procured and stocked in the raceways.

Acclimation and Stocking

While stocking hatchery breeding seeds, a separate acclimation tank may not be required. If necessary, acclimation tanks should be cleaned and disinfected few days, before seeds arrival. Tanks should be filled with filtered water from the same raceway

Figure 6.10: Seeds being Acclimatized

in which the seeds are to be stocked. All water quality instruments *viz.* refractometer, DO meter, pH meter, thermometer, spectrophotometer, etc should be checked and calibrated to ensure accurate reading. Water temperature and salinity should be adjusted according to the expected salinity and temperature of hatchery where seeds are maintained.

To reduce the duration of acclimation, the hatchery personnel (where the seeds are to be purchased) should be asked to commence the salinity adjustment (to attain required ppt) in the larval rearing tank itself. However, if such acclimation is not possible, alternative plans should be arranged to ensure proper acclimation at the nursery raceways. Water temperature, salinity, pH and dissolved oxygen values in the raceway acclimation tanks should be similar to the values in the seed containing bags. To reduce stress, seeds stocking should be done during the cool hours of the day *i.e.* the early morning or late afternoon.

For transport in polythene bags, all master cartons should be unloaded near the raceways. Three to four bags should randomly be selected for water quality analysis. If there are differences (over 5°C/ppt) in water quality between both water sources, gradual acclimation is necessary. The received bags should be allowed to float in the raceway/acclimation tank to adjust temperature. If seeds are transported for more than 10 hrs, it will result in low dissolved oxygen in the transporting bags. In such cases, the bags should be opened and about 2 litres of raceway/acclimation tank water may be added and allowed to float in the raceway/acclimation tank. Since a good algal bloom in the raceway water is expected, special care should be taken to ensure that water pH does not exceed 8.6. Adding water to the bags should be done in increments of 2 l water if the difference between both water sources are minimal. At the end of the acclimation process, seeds can be released into the raceways.

Bottom Condition Observation, Waste Removal and Water Exchange

Raceway bottom condition should be observed every morning before feeding with the help of an underwater viewing apparatus. This apparatus is useful for checking dead shrimp, bottom stagnation, feed leftover and feeding activity. If found, this settled solid wastes will be manually removed by using 1 – inch diameter transparent siphoning hose at weekly once or twice and loss of water in raceway by evaporation and through the process of waste removal can be compensated by adding filtered and U.V. treated water. Fresh water/low saline water may also be added to adjust the salinity.

Table 6.1: Rate of Water Exchange for Raceways

Culture Period (Weeks)	*Daily Water Exchange Rate*
1	< 3 per cent
2	3–5 per cent
3	5–10 per cent
4	10–25 per cent
5	25–50 per cent

The water exchange rate for the first 20 – 25 days can be limited, due to the low biomass and good diatom bloom. The water exchange rate is increased as the shrimp biomass increases. However, the water exchange rate is mainly determined by the water quality parameters and PL health conditions.

The Table 6.1 suggest the daily water exchange rate for nursery rearing period.

A large volume of water can be drained by tilting/detaching swivel pipe in the harvesting tank. To avoid escape of the seeds the outlet filter pipe (fixed inside the raceway tank) is perforated and covered with suitable filter screen (Figure 6.11).

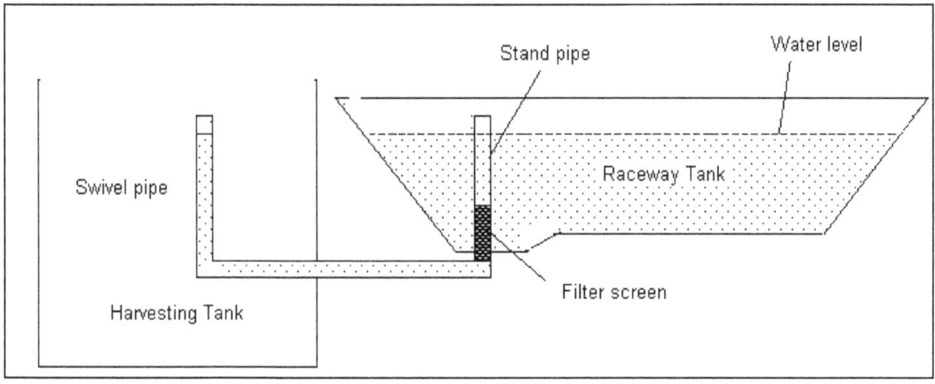

Figure 6.11: Outlet Provision for Raceways

Feeding Management in Raceways

Feeding with dry feed to be started on the first day of stocking. Under - feeding always results in poor growth and increased cannibalism. Over-feeding causes bad water quality and increased cost of production. Hence growth sampling for biomass evaluation should be done weekly once or twice and feed ration should be revised accordingly. Seeds should be fed thrice day with three rations.

Fermented Product Formulation and Preparation for Raceway Aquculture

Fermentation, chemical changes in organic substances produced by the action of enzymes. Zymology, also zymurgy, in biochemistry, study of the processes of fermentation, that is, the chemical changes produced in organic substances by the action of bacteria, enzymes and yeasts. Generally, fermentation results in the breakdown of complex organic substances into simpler ones through the action of catalysis.

Fermented Products to Sustain Microbial Population

Fermented rice bran is used in most shrimp aquaculture systems. The usage of wheat flour, sugar, jaggery and yeast are also used in this aspect. The mass proliferation of heterotrophic bacteria as a feed source in aerated, lined system is a viable option for intensive shrimp production. This can be achieved by manipulating the carbon/nitrogen ratio of culture water through the addition of low protein feeds or molasses. This promotes the formation of bacterial aggregates that are consumed by shrimp. The addition of carbon to increase the carbon/nitrogen ratio in nitrogen rich aquaculture water can reduce toxic ammonia levels in a few hours through the exponential growth of heterotrophic bacteria.

The principle of bioremediation strategy is to use selected microorganisms outside the host to create a healthy environment. It is a biotechnological process of using selected micro or macro organisms to reduce harmful wates to less hazardous levels and in aquaculture, it perhaps provides an effective means to tackle water quality and disease problems in a more natural way using fermented products. Bioremediators improve water quality by nitrification, removal of toxic substances such as ammonia, hydrogen sulphide and nitrite from culture systems. They would efficiently recycle the waste material by decomposing the complex organic substances and generating nutrients in the form of simple inorganic compounds. Of the several strains of bacteria that were tried in the production of an environmental probiotic, gram-positive *Bacillus* spp. are generally more efficient in converting organic matter back to CO_2 than are gram negative bacteria, which would convert a greater percentage of organic carbon to bacterial biomass or slime. Other bacterial species of importance towards bioremediation are: *Nitrobacter, Pseudomonas, Enterobacter, Cellulomonas* and *Rhodopseudomonas* spp. Fermented products used through bioremediators apart from improving water quality parameters, modify or manipulate the inherent microbial communities of water and sediment and reduce the pathogenic microbes to improve the growth and survival of cultured microorganisms. Most of the culture systems are likely to have a build up of ammonia towards the final stages of crop owing to the increased biomass and resultant accumulation of wastes. Nitrifiers in such situations are responsible for the oxidation of ammonia to nitrite and subsequently to nitrate. Addition of nitrifying cultures into such farming systems would be effective in controlling the increasing level of ammonia and nitrites. Such cultures can be developed in the aquaculture system through fermentation technology.

Yeast

The yeast enzyme, zymase, changes the simple sugars into ethanol and carbon dioxide. The fermentation reaction, represented by the simple equation :

$$C_6H_{12}O_6 \rightarrow 2C_2\,H_5OH + 2CO_2$$

It is actually very complex because impure cultures of yeast produce varying amounts of other substances, including fusel oil, glycerin, and various organic acids. Yeast, a microscopic, one-celled fungi important for their ability to ferment carbohydrates in various substances. Yeasts in general are widespread in nature, occurring in the soil and on plants. Most cultivated yeasts belong to the genus Saccharomyces; those known as brewer's yeasts are strains of *S. cerevisiae*. Pure yeast cultures are grown in a medium of sugars, nitrogen sources, minerals and water.

Preparation of Fermented Product for Raceway System

The materials like wheat flour, sugar, jaggery and yeast are used in the fermentation process. The yeast is boiled in the water and then added to the other mixtures prepared in the fermentation can. The required amount of water is added to it. Sufficient aeration should be provided overnight. The fermented mixture is then added to the raceways.

Table 6.2: Composition of an Aquaculture Fermented Product

Sl.No.	*Ingredients*	*Quantity (g/0.1 ha)*
1.	Rice/wheat flour	1000
2.	Sugar	1000
3.	Yeast (activated by dissolving in warm water and inoculated)	10
4.	Pond water	25 litres

Figure 6.12: Fermentation Chamber

Algal Culture Concentrate–Production for Raceways

Refer another chapter for details.

Algal Culture Operations and Methods

Isolation of Pure Algal Strain from Raw Sea Water

Twenty liters of sea water collected during high tide should be enriched with nutrients and left under light till algae bloom. The nutrients added to the sea water are suitable for the growth of algal species and favour their dominance in the culture. Hence, sub culturing should be repeated till *Chaetoceros* or other required algae dominates in the bloom. The pure strain can be obtained from the bloom by serial dilution. Pure strain should be cultured in 50 ml sterilized flasks with enriched sea water till a density of one million cells per ml is obtained. This is the stock or starter for mass culture.

Preservation of Stock Culture

Small flasks (50ml) or test tubes (20ml) filled with enriched sea water are inoculated with 0.1 ml of stock culture and incubated in light with a photo-period of 12 hrs. In this method, the algae can be stored and maintained for 15 days. Afterwards, the above procedure should be repeated to keep the algae in the active growth phase. The stock culture can be stored in refrigerator for one month.

Flask Culture

The stock culture maintained in 50 ml flasks passes through many progressive culture steps before it reaches mass culture phase. Twenty ml of stock culture is inoculated into 250 ml sterile flasks with enriched sea water and incubated in light for two days with continuous aeration, to get a density of 5 million cells per ml. Some of these 250 ml flask will be used for inoculating the small flasks. After 2 days, these 250ml flasks should be transferred to 2 liter flasks with enriched sea water and incubated in light with aeration for 2 days to get 3 million cells per ml density. This is inoculated into 20 liter carboys with enriched sea water and incubated in light with aeration for 2 days to get a density of 3 million cells per ml.

Figure 6.13: Indoor Algal Culture Facility

Once in 15 days a new starter culture will be introduced into the system to maintain the vigor of the culture.

Mass Culture

The carboy culture of 3 million cells per ml density will be transferred into FRP cylinders (200 lt.) filled with enriched sea water. The cylinders are aerated and incubated in light for 3 days. When it reaches 1 million cells per ml density it is transferred to the larval rearing tanks. Where there is no facility for indoor mass culture, indoor culture can be stopped at 2 liter flasks level. And outdoor mass culture can be started with inoculating 20 liter buckets with 2 liter flask culture. The buckets should be incubated in sunlight with aeration for 2 days till the density reaches 3 million cells per ml. Then the culture is transferred to 100 liter cans with enriched seawater which are then aerated and incubated in sunlight for 2 days to get a density of 2 millions cells fiber glass tanks (1MT) filled with enriched sea water. The tanks should be aerated and incubated in million cells per ml. Then the culture is transferred to larval rearing tanks.

Algal culture operations must be started, several days prior to the acquisition of the gravid shrimp, taking into consideration the time taken for isolation and mass culture, so that the mass culture reaches peak growth just before the larvae begin feeding.

In mass culture, algae will be in the first phase on the first day of inoculation and for the next two days (2nd and 3rd day) it will be in exponential phase. On the fourth day it reaches the declining phase of relative growth, then stationary and death phase. Prolonged culture results in a decrease in size and nutritive value.

Harvest and Feeding

The algal water from indoor and outdoor mass culture tanks is pumped directly to the larval tanks in desirable volume using a submersible pump. The pump and pipe should be disinfected and washed thoroughly before and after use.

Do's and Dont's in Raceways

Do's

1. Setting up of a biosecurity program that looks after the specific needs of the culture system.
2. Ensure turbulent aeration with zero water exchange.
3. Evaluation of healthy or diseased free stock from hatcheries.
4. Proper acclimation and transportation of shrimps seeds.
5. Adequacy and efficiency of filteration systems.
6. Maintain the water quality parameters within the limits accurately.
7. Exchange the water whenever necessary and always drain from the bottom of the tanks to help flush sediments accumulated near outlets.
8. Ensure overall feed quality.
9. Routine sampling to examine the health and growth of shrimps. During serious disease outbreaks that spread fast like WSSV and YHV, harvest the remaining shrimp, disinfect the raceways system without discharging any water, prevent the escape of animals into the environment.
10. Raceways should be cleaned and disinfected between crops.

11. Dead shrimp from raceway systems should be disposed in a sanitary manner, such as treating with quicklime followed by burial at 60 – 90 cm depth.
12. Monitoring the raceways during night hours is paramount important.
13. Continuous running of blower is essential. Hence in case of power failures the generator back-up is inevitable.
14. Adequate storage of feed, algae, fermented product and fuel needs to be ensured.
15. A spare blower will be handy to run the raceways without interruptions.

Dont's

1. Abrupt light or loud noise interruption should be avoided to prevent the post larvae from jumping out of raceway tanks.
2. Unfavourble water quality conditions should be avoided which will lead the larvae to jump or swim close to the water surface during the day time.
3. Anoxic condition on the bottom of the raceway should be carefully avoided.
4. Avoid left over feed on the raceway system.
5. High pH above 9.0 should be avoided which will cause high algal densities; Low pH of below 7 will lead to heavy bacterial load and high shrimp biomass.
6. Too high water temperature and too low unicellular algal bloom will make the system flourish with blue green algae (BGA).
7. Algal crash should be avoided which will be caused by high ammonia level.
8. Blockage in air lift pump by algae or other objects should be avoided.
9. The high pumping rate of air-lift pump should be avoided.
10. Raceways turning transparent needs to be avoided.
11. Excess air flow into the raceways is not advisable.
12. Avoid using additional inputs like probiotics of unknown origin (commercial) in the raceways.
13. Do not stock seeds without PCR screening.
14. Do not forget to repeat the PCR run throughout the crop.
15. Routine maintenance of all farm equipments should not be neglected.

References

APHA. 1985. Standard methods for the examination of water and waste water, 15th Edn. American Public Water Works Association and Water Pollution Control Federation, New York.

Felix, S. 2011. "Innovative culture systems to enhance inland fish production in India". *Info fish International Journal* Vol. 4; 31-34.

Felix.S., 2012. "Bioscured aerobic Flocculent Driven Raceway Farming Technology for Mass Producing Ornamental Fishes in India". Proceeding of International Conference on "Sustainable Ornamental Fisheries- Way Forward" at Cochin University of Science and Technology, Kochi from 23rd to 25th of March 2012 pp. - 200.

Hargreaves JA, Kucuk. S. 2001. Effects of diel un-ionized ammonia fluctuation on juvenile hybrid striped bass, channel catfish, and blue tilapia. Aquaculture, 195: 163-181.

Mirzoyan, N., Tal, Y., Gross, A. 2010. "Anaerobic digestion of sludge from intensive recirculating aquaculture systems: Review" In: "Aquaculture" 306 (2010) 1–6.

Yoram Avnimelech. 2011. Tilapia Production Using Biofloc Technology Saving Water, Waste Recycling Improves Economics. Global Aquaculture Advocate. No. 1:66 (66-68).

Part IV

Genetically Modified Organisms

Chapter 7

Genetically Modified Organisms in Aquarium Trade

B. Ahilan

Department of Aquaculture,
Fisheries College and Research Institute,
Thoothukudi – 628 008, T.N.

Genetic engineering processes are becoming increasingly common and these are being applied to a widening variety of organisms. It has brought with it revolutions in the life sciences over the past decade. The beneficial trait of a particular gene from organism can not be transferred to another in a less amount of time and with new precision. This technology has already achieved a tremendous success in agriculture and made it easier for farmers to grow certain crops and reduced the use of certain pesticides. In recent years, genetic engineering has also been applied to produce specific animal strains generally referred to as Genetically Modified Organisms (GMOs) that also have the potential to improve live stock crops. Transgenic technology is widely used in biotechnology, from generation of genetically modified (GM) feeds to production of pharmaceutical protein. However, its commercial application needs careful evaluation to address potential environmental, social and regulatory concerns. Indeed, the first commercially marketed, genetically engineered animal in the United States of America (USA), and several other countries, was for the recreational hobby and the aquarium, market the Glo fish.

Genetically Modified Organisms

An organism that has a foreign or modified gene integrated in its genome using the in vitro genetic techniques of genetic engineering is called as "transgenic" or "genetically modified organism" (GMO). The GMOs are also referred as "Living

Modified Organism" (LMO). It means any living organism that possesses a novel combination of genetic material obtained through the use of modern biotechnology.

The synonymous of GMO are LMO, Transgenic and GEO (Genetically Engineered or Enhanced or Enriched Organism). For the production of transgenic animal, a cloned gene is transferred into the fertilized egg before the cleavage.

Why GMOs are Produced?

Genetic engineering is altering the genetics of a variety of species for the purpose of increasing aquaculture production, medical, cleaning up water pollutants and ornamental reasons

Genetic engineering is commercial feasible way of bridging large gaps between an organism's natural characteristics and what the aqua culturists wants.

The main reasons for genetic manipulation of species used in aquaculture are directly connected to improved output/input ratios. The areas are as follows:

1. To enhance growth and or efficiency of feed conversion
2. To increase tolerance to environmental variables such as temperature and salinity
3. To produce new colour variants of ornamental species
4. To enhance commercially significant flesh characteristics
5. To control reproductive activity
6. To increase resistance of species to pathogens/parasites
7. To produce novel medicinal substances with fewer animals welfare problems than when mammals are used.

Aquatic GMOs

Application of biotechnology to aquatic and marine organisms has been an area of dramatic technical advancement. Transgenic technology in fish has come a long way since 1985, when the first successful transgenic fish was reported by Zhu *et al.*, Currently numerous laboratories around the world are working on transgenic fish, resulting in more than 3000 publications related to transgenic fish. In the 1980s and early 1990s, the impetus to develop transgenic fish was largely driven by the interest of generating superior fish stocks for aquaculture. This has led to the development of fish with useful traits in aquaculture. Till 1995, most transgenic fish research was focused on the establishment of a transgenic fish model for developmental analyses.

At least 35 species of transgenic fish are currently being developed around the world. Since then, many species have been used to produce GMOs as shown in Table 7.1.

1. Transgenic Fish Production/Aquatic GMOs Production

Production of GMOs is a multistage process, which can be summarized as follows:

- ☆ Identification of genes of interest

Table 7.1: Some Aquatic GMO's being Tested for Use in Aquaculture

Species	*Foreign Gene*	*Desired Effect and Comments*	*Country*
Atlantic salmon (Davies *et al.*, 1989b, Fletcher *et al.*, 1988, 1992 and 2004, Hew *et al.*, 1992 and 1995, Du *et al.*, 1992)	AFPAFP Salmon GH	✰ Cold tolerance ✰ Increased growth ✰ Feed efficiency	United States, Canada
Coho salmon (Delvin *et al.*, 1995)	Chinook Salmon GH + AFP	✰ After 1 year, 10 to 30 fold growth increase	Canada
Chinook salmon (Sin *et al.*, 1993)	AFP salmon GH	✰ Increased growth feed efficiency	New Zealand
Rainbow trout (Chourrout *et al.*, 1986, Nilsson *et al.*, 1992, Agellon *et al.*, 1988, Disney 1988)	AFP salmon GH	✰ Increased growth ✰ Feed efficiency	United states, Canada
Cutthroat trout (Delvin *et al.*, 1995)	Chinook salmon GH + AFP	Increased growth	Canada
Tilapia (Brem *et al.*, 1988, Moclean *et al.*, 1995 and 2002, Moatinez *et al.*, 1996, Rahman *et al.*, 1998 and 2000)	AFP Salmon GH	Increased growth and stable inheritance	Cuba
	Tilapia GH	Increased growth and stable inheritance	Cuba
Tilapia	Modified tilapia insulin – producing gene	Production of human insulin for diabetics	Canada
Salmon (Hew *et al.*, 1995)	Rainbow trout lysosome gene and flounder Pleurocidin gene	Disease resistance, still in development	United State
Striped bass	Insect genes	Disease resistance, still in early stages of research	United States
Mud loach (Zhu *et al.*, 1986, Tsai *et al.*, 1995)	Mud loach GH + mud loach and mouse promoter genes	Increased growth and feed efficiency, 2 to 30 fold increase in growth, inheritable transgene	China and Korea, Rep.
Channel catfish (Dunham *et al.*, 1992)	GH	33 per cent growth improvement in culture conditions	United States
Common carp (Zhu *et al.*, 1989, Zhang *et al.*, 1990, Chen *et al.*, 1993, Chatakondi *et al.*, 1995)	Salmon and human GH	150 per cent growth improvement in culture conditions; improved disease resistance; tolerance of low oxygen level.	China and United States
Indian major carps (Venugopal *et al.*, 2002a)	Human GH	Increased growth	India

Contd...

Table 7.1–*Contd...*

Species	*Foreign Gene*	*Desired Effect and Comments*	*Country*
Goldfish (Zhu *et al.*, 1985)	GH AFP	Increased growth	China
Northern Pike (Gross *et al.*, 1992)	GH	Increased growth	United States
Abalone	Coho Salmon GH + Various promoters	Increased growth	United States
Oysters	Coho Salmon GH + Various promoter	Increased growth	United States

- ☆ Isolation of these specific genes
- ☆ Amplifying the gene to produce many copies
- ☆ Associating the gene with an appropriate promoter and poly – A sequence and insertion into plasmid
- ☆ Multiplying the plasmid in bacteria and recovering the cloned construct for injection.
- ☆ Transference of he construct into the recipient tissue, usually fertilized eggs.
- ☆ Integration of gene into recipient genome
- ☆ Expression of gene in recipient genome and
- ☆ Inheritance of gene through further generators

1.1 Selection of Fish Species

For their aquaculture importance, Indian major carps, Common carp, Channel catfish, Chinese carps, Salmon, Trout are the strong candidate species for the transgenic project. Small aquarium fishes like Zebra fish and Japanese medaka are most suitable because they (i) breed through out the year in the laboratory (ii) lay a large number of eggs at a time and (iii) possess short generation time (*i.e.* they mature at about 3 months of age).

1.2. Selection of Target Gene

i) Growth Hormone Gene

The most popular gene used in aquatic species is growth hormone (GH) for reasons that are obvious. Transgenic fish carrying GH gene, will produce growth hormone endogenously by passing the necessity of exogenous hormone treatment. GH gene has been cloned in some fishes either from the genomic library or from cDNA library.

ii) Disease Resistance Gene

The potential of Rainbow trout lysozyme gene as a bacterial inhibitor was assessed in Atlantic salmon.

iii) Cold Tolerance Gene

Some marine teleosts have high levels of serum Anti-Freeze Proteins (AFP) or Glycoproteins (AFGP), which reduce the freezing temperature by preventing ice – crystal growth. The production of AFP from two species of winter flounder (*Pleuronectes americanus, Pseudopleuronectes amenis)* have been achieved.

1.3. Transgene Delivery

There are several methods of transferring the gene into the animal embryo, which include i) Microinjection ii) Electroporation iii) Use of retroviral vector vi) Lipofection and v) Use of embryonic stem cells.

Transgenic fish have largely been produced through microinjection into fertilized eggs or early embryos. It is a tedious and slow procedure and can result in fish egg mortality.

Electroporation involves placing the eggs in a buffer solution containing DNA and applying short electrical pulses to theoretically create a transient opening of the cell membrane, allowing the transfer of genetic material from solution into cells. Electroporation can be more efficient than microinjection with integration rates sometimes as high as 30 – 100 per cent.

Use of pantropic retroviral vector is able to infect a wide range of host cells and have been Medaka eggs. The use of viral promoters should be avoided for producing transgenic feed fishes, as some of these viruses are reported to induce cancer formation. Liposome's have also been utilized as vectors. Here uptake of nucleic acid encapsulated with synthetic lipid vesicles is permitted in to the cells following the fusion of the vesicles with the plasma membrane.

Recently the use of embryonic stem cells (ESC) as a method for inducing transgenesis has been advocated. Similarity primordial germ cell (PGC) transplantation technique has also been developed for fish, providing another powerful approach for the production of transgenic fish.

Efficiency of gene transfer is determined by several factors including hatching percentage, gene integration frequency, the number of eggs which can be manipulated in a given amount of time and the quantity of effort required to manipulate the embryo in this regard. Electroporation is a powerful technique for mass production of transgenic fish. Integration, expression and transmission of the pantropic retroviral reporter transgene observed for at least three generations. This is a very good gene transfer technique for live – bearers and fish in general, but introduction of viral sequences, into feed fish may not be accepted by the public.

2. Development of Genetically Modified Organisms for Ornamental Fish

Growth enhancement for human food production in aquaculture is the most common objective of current efforts but it may not remain so for much longer. Indeed, the first commercially marketed, genetically engineered animal in USA and several other countries was for the recreational hobby aquarium, market the Glo fish, a transgenic fish that glows due to skeletal muscle expression of a fluorescent protein genetic construct.

The increasing world demand for ornamental fish are opened the market for new varieties with novel shapes or colours, which can be supplied through the use of transgenics.

2.1. Genetically Engineered Fluorescent Zebra Fish

Scientists develop the transgenic fluorescent zebra fish by first identifying and isolating the fluorescent protein gene in marine organisms (such as jelly fish or sea anemone). The isolated natural fluorescence genes are introduced into the fish eggs before they hatch. Once the gene integrates into the fish's genome, it will be able to pass the fluorescence gene on to its offspring. The transgenic fluorescent zebra fish are similar to other normal zebra fish every way except for their brilliant colours and flourscence. The fluorescent fish are available in star-fire red, sun – burst yellow and electric green. Production of novel coloured fish required fusion of florescent colour-encoding genes to appropriate tissue specific promoters. So far, expression of fluorescent protein has been directed to the skin or several muscles, allowing colourful fish to be visualized under normal daylight. In addition two – colour fishes also produced, showing one colouration in the skin and another one in skeletal muscles. Such an approach, combined with selective breeding between fish carrying these different transgenics, could produce a wide array of multi coloured fish in future generations.

Flourescent protein genes can be introduced into commonly available and less expensive species such as *Puntius* spp., *Mystus* spp., and glass fishes for value addition and also in carp and gold fish.

2.1.1 Fluorescent Protein Gene

These genes are capable of producting the beautifully colured florescent protein. The Green Flourescent Protein (GFP) originally isolated from the jelly fish (*Aequorea victorea)* is intrinsically fluorescent allowing direct visualization without the need of substrate for chemical reaction. GFP is a valuable tool in the colour tansgenics. It has many advantages.

1. It does not require substrate over it to be irradiated by blue light so, not limited by availability of the substrate;
2. The gene expression can be monitored in living cells/animals.
3. The substance is not toxic to living cells/animals
4. The protein is persistent even in formaldehyde fixed tissues.

The coding DNA (cDNA) for this protein has been cloned and modified by site – directed mutagenesis for different emission spectra and thus several artificial fluorescent colour proteins become available, including Yelow Fluorescent Protein (YFP), Blue Fluorescent Protein (BFP) and Cyano Fluorescent Protein (CFP). More recently, a new fluorescent protein cDNA encoding a Red Fluorescent Protein (RFP) has been cloned form the Indo pacific sea anemone relative (*Discosoma* SP.). Due to the fact the these fluorescent proteins can be observed in live biological samples for labeling and subcellular organelles, these fluorescent proteins have been aptly termed "Living colours".

2.1.2 Mechanism of Fluorescence

Fluorecscence is a phenomenon in which energy form a source of light is absorbed and re-emitted as another photon. The florescent fish absorb light and then re-emit it and thus appear to be glowing, particularly in a dark room. In fluorescence, the mechanism is different. The ground state electrons are present in lower energy orbital. Once the light (photon) strikes the fluorescent protein, the electrons in the ground state become excited and jump to higher energy orbital. The electrons is only stable there for a short time, if then returns to the lower energy level by emitting energy as a longer wave length photon. This longer wave length photon is responsible for the fluorescence phenomena.

2.1.3. Application of Glow Fish

1. As Ornamental Fish

The fluorescent zebra fish is excellent example of genetically engineered ornamental fish. Fluorescence makes it a unique ornamental fish. The fluorescent zebra fish is a typical tropical fresh water ornamental fish. The fish increase the beauty of aquaria especially during the night by emitting light, so referred as "Night pearl". It fulfills almost all the requirements of an ideal ornamental fish.

2. As Pollution Indicators

The transgenic zebra fish can detect water pollution by changing colour. Scientists have developed commercially viable zebra fish that can be used as pollution indicators making them a simple alternative to a complicated pollution testing system.

The glowing zebra fish can be used to identify pollutants in drinking water. The fish light up when exposed to Polychlorinated Biphenyls (PCBs), which are known to cause cancer in human beings. These fish are much more sensitive than current water testing systems and can detect low concentrations of PCBs. Testing water using fluorescent zebra fish takes less time and cost than testing with conventional equipment.

Normal zebra fish are usually black and silver in colour, which transgenic zebra fish are capable of producing green or red fluorescent colours. The fluorescence in fish is under the control of genes. In order to trigger off the genes in the fish to be of any use, inducible gene promoters are used to act as control switches to activate different tissues of the fish. At present, scientists have succeeded in isolating two types of gene promoters in the zebra fish: an estrogen – inducible promoter and a stress-responsible promoter.

The promoters have been used to drive the fluorescent colour genes in transgenic zebra fish. Such fluorescent coloured transgenic fish will be able to respond to the presence of chemicals such as estrogen through the estrogenic promoter whereas heavy metals and toxins respond through the estrogenic promoter whereas heavy metals and toxicants. The fish will display the fluorescent colour that has been assigned to detect the particular contaminants. The detection of pollutants by transgenic fish is very quack and economical too. Furthermore, these fish are also biodegradable. All these factors make them very suitable pollutant indicators.

Scientists are also working to provide a fish that displays different coloured glows depending on the surrounding water temperature. This may lead to the use of fluorescent fish as live thermometers (temperature indicators).

3. As Experimental Organisms

Scientists to seek answers to important questions in molecular biology, genetics and vertebrate development also use zebra fish. Fluorescent zebra fish have been particularly helpful in increasing our understanding of cellular diseases and developmental biology as well as in cancer research and gene therapy.

Risk Factors of GMOs

Potentially useful transgenic fish strains have been developed in different parts of the world, but their widespread use in aquaculture currently has not yet occurred because of social issues and the concerns associated with the potential effects that GM fish may have eon natural ecosystems. The most important areas of risks, which need to be considered in the use of transgenics are biodiversity, human health and animal welfare.

Aquatic GMOs Research in India

Research on animal transgenesis was initiated by National Institute of Immunology (NII) in India. Thereafter, investigations have been made on rohu (*Labeo rohita),* zebra *(Brachydanio rerio),* and catfish. The first Indian transgenic zebrafish was generated in 1991, followed by first triploid transgenic *Brochydanio rerio* in 1995, using borrowed constructs from foreign sources. At centre for Cellular and Molecular Biology (CCMB), Hyderabad, autotransgenic Catla *catla* and *Labeo rohita* were generated using the growth hormone gene constructs (both cDNA and genomic DNA) fully developed from these species. Efforts are on way to isolate and characterize salt – resistant genes from marine environment in a collaborative project between Central Institute of Fisheries Education (CIFE) Mumbai and CCMB, Hyderabad. Similarly, attempts have been initiated at M.S. Swaminathan Research Foundation, Chennai, Central Marine Fisheries Resources (NBFGR) and Central Institute of Fisheries Technology (CIFT), Kochi to isolate and characterize salt tolerance genes from mangrove plants like *Avicenia* sp., sea grasses and marine microbes. Indian Council of Agricultural Research (ICAR) has recently initiated research on transgenic fish with an approach to increase growth rate and to produce biosensors.

GMOs in Aquarium Trade

Glo Fish

The Glo fish was introduced to the United States market in December, 2003 by Yorktown Technologies of Austin, Texas, after more than two years of extensive environmental research and consultation with various Federal and state agencies, as well as leading experts in the field of risk assessment was made by the U.S. Feed and Drug Administration. The Glo fish is a trade market brand of GM transgenic fluorescent zebra fish (*Danio rerio).*

1. Commercialization of Glofish in the United States poses regulatory uncertainly because existing biotechnology policy bases oversight on the use of the product.
2. Sales of ornamental fishes are not federally regulated. The feed and Drug Administration asserts jurisdiction over genetically modified animals using the New Animal Drug Application Process.

FDA allowed commercialization of Glofish without regulating approval. FDA announced: "Because tropical aquarium fish are not used for feed purposes they pose not threat to the food supply. There is no evidence that these GE *Zebra danio* fish pose any more threat to the environment than their unmodified counter parts, which have long been widely sold in the United States. In the absence of a clear risk to the public health, the FDA finds no reasons to regulate these particular fish".

Marketing of the fish was met by protests from a Non-Governmental organization called the Centre for Feed safety (CFS). They were concerned that approval of the Glo fish based only on Feed and Drug Administration risk assessment would create a precedent of inadequate scruting of biotech animals in general. In January 2004, CFS filed a law suit against the FDA.

The CFS's suit was dismissed on March, 30, 2005.

Developments during 2006 - 2007

Glofish have continued to be successfully marketed throughout the United States. After more than three years of availability, there are no reports of any ecological concerns associated with their sale.

In addition to the red fluorescent zebra fish, trade marked as "Star fire Red", York Town Technologies released a green fluorescent zebra fish and an orange fluorescent zebra fish in mid – 2006. the New lines of fish are trademarked as "Electric Green" and "Sunburst Orange", and incorporate genes from sea coral.

As of January 2007, sale or possession of Glofish is illegal in California due to a regulation that restricts all genetically modified fish. The regulation was implemented before the marketing of Glofish, largely due to concern about a fast growing biotech salmon. Canada also prohibits import or sale of the fish, due to what they report is a lack of sufficient information to make a decision with regard to safety.

Marketing of fluorescent zebra fish may spur efforts to develop other transgenic ornamentals such as gold fish and koi carp.

Future Perspectives

Currently, there is not a great amount of research on transgenic fish throughout the world. However, there is tremendous worldwide investment in genomic research.

One of the greatest future potential benefits of gene transfer in fish will be enhancement of disease resistance in fish. Transgenic fish with enhanced disease resistance would increase profitability, production, efficiency and the welfare of the cultured fish.

Transfer of growth hormone gene constructs has dramatically altered growth rates of fish. Once application of these fish are allowed, major impacts on aquaculture production can be expected. It is also apparent that transgenics alter, the body composition. Data to data indicates that this has the potential for future positive impact.

Apparently, one of first application of transgenic fish will be alteration of colour or ornamentation and aquarium fish with fluorescent pigment genes consumers have a demand for these colour altered transgenic fish, this could evolve into a major application of transgenesis with a large economic impact. This may also result in additional environmental fish issues and confinement issues. But not all ornamentals can survive in the natural environment.

Biostatic issues of transgenic fish also need very careful attention. The future success and application of transgenic fish will be dictated by successful demonstration of a lack or potential lack of environmental risk, feed safety, appropriate government regulation and development of genetic sterilization for transgenic fish.

Conclusion

GMO in aquaculture have much to offer in terms of improvements in fish production, feed security and generating economic benefits. Transgenic research related to modified fish for ornamental trade and other industrial products and as biosensors to monitor water pollution levels are more attractive and with controversies. Due attention also to be paid on research (i) to generate new varieties of expressive endemic ornamental fish varieties and (ii) for value addition in commonly available and less expensive other native species such as *Puntius Spp., Mystus Spp.,* and glass fishes. As demonstrated by the current plan to regulate transgenic fish, it is unclear if regulators have the tools they need to adequately evaluate these new products, and there is reason to wonder if an innovation is getting ahead of our ability to manage it. With out clearly articulated and transparent road map to guide agency review and approval, it will remain difficult for developers to bring products to market.

References

Ayyappan, S. and Gapalakrishnan, A. 2006. Transgenics in fisheries: Perspective, priorities and preparedness for India. *Indian J. Fish.* 53 (2):127-152.

Beaumont, A.R. and Hoare, K. 2003. Biotechnology and Genetics in Fisheries and Aquaculture. Black well publishing, UK. pp: 126 – 140.

Chakraborty,C. and Haques,s. 2000. Edible vaccine from Transgenic fish. *Fishing chimes.* 20: 104, 105.

Chandra, G. and Ahmed, I. 2007. Genetically engineered glow fish – fluorescent beauties with practical application. *INFOFISH international.* 5: 22 – 25.

Dunharn, R.A. 2004. Aquaculture and Fisheries Biotechnology Genetic Approaches. CABI publishing, USA. pp: 160 – 243.

Gong, Z. and Korzh, V. 2004. Fish development and Genetics, world scientific, London. pp: 476 – 516.

Gong, Z., Wan, H. and Yan, T.L. 2003. Development of transgenic fish for ornamental and bioreactor by strong expression of fluorescent proteins in the skeletal muscle. *Biochemical and Biophysical Research Communications.* 308 : 58 – 63.

Kapuscinski, A.R. 2005. Current scientific understanding of the environmental biostatic of transgenic fish and shell fish. *Rev. sci.tech.off int. Epiz.* 24(1): 309 – 322.

Karthik, M., Akolkar, D.B., Kumar, V. and Singh, S.D. 2006. Genetically modified fish and its potential applications. *Aquaculture Asia magazine.* April-june :13 – 17.

Kinoshita, M. 2004. Transgenic Medaka with brilliant fluorescence in skeletal muscle under normal light. *Fisheries science*. 70: 645 – 649.

Kumar, V., Vijayan, H. and Saurabh, S. 2005. Green fluorescent protein from jelly fish, *Aequorea victoria* and its promising applications in molecular biology. *Aqua international*. September: 26-31.

Maclean, N. and Laight, R.J. 2000. Transgenic fish: an evaluation of benefits and risks. *Fish and fisheries*. 1: 146 – 172.

Pandian, T.J. 2001. Guidelines for research and utilization of genetically modified fish. *Current Science*. 82 : 1172 – 1178.

Pandian,T.J., kavumpurath.,S., Mathavan,S. and Dharmalingam. 1991. Microinjection of rat growth hormone gene into zebra fish egg and production of transgenic zebrafish. *Current science*. 60 : 596 – 600.

Rahman, M.A. and Maclean, N. 1999. Growth performance of transgenic tilapia contaminating an exogenous piscine growth hormone gene. *Aquaculture*. 73:333 – 346.

Reddy, P.V.G.K., Ayyappan, S., Thampy, D.M. and Krishna, G. 2005. Text Book of Fish Genetics and Biotechnology. Indian Council of Agricultural Research, New Delhi. pp: 65 – 86.

Sachin, B., Starm, S.R., Yadav. and Desai, A.S. 2007. Threats and Regulatory measures in Transgenic fish culture. *The Asian Journal of Animal Science*. 2: 109 – 113.

Singh, B. 2006. Marine Biotechnology and Aquaculture development. Vista International Publishing House, Delhi. pp: 137 – 147.

Thomas, P.C., Suresh, R., Mohapatra, K.D. and Pillay, T.V.R. 2003. Breeding and seed production of finfish and shellfish. Daya Publishing House, Delhi. pp: 295 – 30.

Part V

Natural Colour Development in Ornamental Fish

Chapter 8

Role of Pigments for Enhancing Colours in Ornamental Fishes

Saroj K. Swain

Central Institute of Freshwater Aquaculture,
(Indian Council of Agricultural Research),
Kausalyaganga, Bhubaneswar – 751 002

Ornamental fishes fetches good price if the colour is bright and vibrant. Scientists are presently exploring the colour enhancers from various sources. Colour enhancement through the use of carotenoids in feed is now being examined in various organizations although very few scientific papers were published on this topic. Carotenoids (pigments) are the primary source of colour in the skin of fish. The various colours are produced by the presence of specific carotenoids and carotenoid-protein complexes. Since carotenoids are only synthesized by plants and can be utilized in animal tissue when fishes are kept in fertile ponds, they obtain carotenoids by eating aquatic plants and small fauna in the aquatic feed chain and enhance their natural colour. However, with the increase of modern intensive re-circulating systems and the variability in the productivity of ponds, the amount of carotenoids available may not always be sufficient to give satisfactory skin colour. Therefore there is a need to study in detail about the various pigments responsible for ornamental fishes and their successful inclusion in the diets.

Carotenoids

Carotenoids are a group of over 600 natural lipid-soluble pigments that are primarily produced within phytoplankton, algae and plants. These pigments are responsible for the broad variety of colors in nature; most notable are the yellow, orange and red colours of fruits, leaves and aquatic animals. Among all these numerous classes of natural colors, the carotenoids are the most widespread and structurally diverse pigmenting agents. Although plants, algae and some fungal and bacterial species synthesize carotenoids, animals cannot produce them *de novo*. Carotenoids are absorbed in animal diets, sometimes transformed into other carotenoids, and incorporated into various tissues. For example, Flamingos ingest algae containing high levels of beta-carotene and convert this yellow carotenoid into canthaxanthin and astaxanthin before depositing it into the feathers and tissues as red plumage. Some fish species such as koi and various crustaceans (*P. japonicus* and *P. monodon*) have the enzymatic mechanisms to convert carotenoids into other forms such as astaxanthin.

Natural Carotenoids

The largest groups are the fat soluble carotenes and xanthophylls. The carotenes are found mainly in green plant material and have Alpha and Beta forms. B-carotene is the precursor for vitamin A. This conversion is carried out universally by most animals. The xanthophylls are synthesised in green material starting during early summer and maturing to the yellow/red pigments (fruits, flowers, roots). Examples are tomatoes, maize, carrots, beetroot and mushrooms. Natural carotenoids are to be found in almost all sections of the living world. Without the pigments perhaps the world would not have been so colourful and nature would not function as we know it. Carotenoids have roles in reproduction (Breeding), respiration (Chlorophyll pigments), light absorption, reflection and the efficiency of the immune system.

Pigment for Other Use

Many animals use colour to communicate warnings, mating calls, feeding signals and camouflage. Fishes also display their body colours for multiple purposes They are communication, identification, camouflage, defense, mimicry and adoption to the environment.

Carotenoids in Aquatic Environment

The predominant carotenoids present in the aquatic environment are specifically lutein (greenish-yellow), zeaxanthin (yellow- orange), alpha-carotene (light yellow), astaxanthin (red) is found in aquatic animals in greatest abundance, The main sources being molluscs, crustaceans and fish particularly shrimps, prawns,trout and salmon. Carotenoids in the wider animal kingdom are found in various insects, butter, egg yolk and cheese. Besides natural sources, chemically synthesised forms of individual carotenoids are available in gelatine protected beadlets. These are astaxanthin (roche carophyll pink), canthaxanthin, citranaxanthin. Astaxanthin is the main commercially synthesized carotenoid used in koi feeds. Although chemically synthesized products have the main advantage of being more reliable in composition and stability, they do not contribute to the range of carotenoids to be found in natural products. Some of the field work conducted at ornamental fish breeding unit of CIFA indicates that a number of ornamental fish species fed natural sources like zooplankton, tubifex, blood worm, etc besides fish raised in green water containing algae had better colouration than when fed synthetic pigments like beta carotene.

The reason is natural form is being preferentially absorbed quickly and/or contributing a number of additional carotenoids with beneficial effects. Shrimp, krill, prawn and spirulina are examples of raw materials that have this beneficial effect.

Carotenoids in Plants

The natural extracts first used were those of carrot, palm oil, saffron, tomato and paparika (Britton, 1996). The powdered, dried plant materials and extracts of them have been used for many years. Generally green leaves contain large amount of carotenoid. Carotenoids are often considered to be plant pigments alone but also occur widely in bacteria, fungi, algae and in animals. Goodwin (1980) provided extensive details about the distribution about carotenoids in plants and animals.

In plants, the carotenoids occur universally in the chloroplast of green tissues, but their colour is masked by the chlorophylls. The leaves of virtually cell species contain the same main carotenoids, that is β-carotene (usually 25-30 per cent of the total), lutein (around 45 per cent), violaxanthin (15 per cent) and neoxanthin (15 per cent) (Britton,1996). Small amount of β-carotene,cryptoxanthin, zeaxanthin, antheraxanthin and lutein 5, 6 epoxide are also frequently present (Cogdwell, 1988).

Carotenoids are also widely distributed in non-photosynthetic tissues of plants and are responsible for the yellow, orange and red colour of many flowers and fruit. They are usually located in chromoplasts (Goodwin, 1980). Natural extracts of some flower such as marigold (Tagetes erecta) contain large amount of lutein and are used for colouration.

Total carotenoid content varied from species to species. Chandersekar *et al.* (1999) reported, the carotenoid content in amaranthus, coriander and mint and the levels were found to be 37,400 g/100 g, 15,077 g/100 g and 8,305 g/g respectively in raw form. They used spectrophotometry and high performance liquid chromatography for the analysis. Kowsalya *et al.* (2001) reported that the total carotenoid content of these were 29.064 mg/100 g, 36.720 mg/100 g and 29.057 mg/100 g respectively on the basis of dry weight (sun dried).

Amaranths are hardy, wild, uncultivated, fast growing cereal-like plants. The growth habits vary from horizontal to erect and branched to unbranched; their leaves and stems are of many shades of green and seeds was from black to white. Amaranth is one of the few double duty plants which can supply tasty leafy vegetables, as well as grains of high nutritional quality. The leaves of all 50-60 species of amaranths are edible (Singhal and Kulkarni, 1988).

Experiments in Ornamental Fishes

There are few research work have been made for determining colouration in ornamental fishes. The reason for enhancing colour is well defined in aquatic syestem. Ako *et al.* (1999) observed a colour enhancement in fresh water red velvet swordtail (*Xiphophorus helleri*), rainbow fish (*Pseudomugil turcatus*) and Topaz cichlids (*Cichlasoma myrnae*) by using diets containing *Spirulina platensis* (1.5-2 per cent) and 1 per cent *Haematococus pluvialis*.

Gouveia, *et al.* (2002) developed coloured ornamental fish (*Cyprinus carpio* and *Carassius auratus*) by micro algal biomass supplementation.

Different type of Carotenoids and their concentration effect in commercially formulated fish diets on colour and its development in the skin of the red oranda variety of goldfish were reported (Geoffrey *et al.*, 2005).

Sherief and Mathew (1996) observed colour enhancement in goldfish using different ingredients such as carrot meal, beetroot meal and red grape skin meal. Lovell (1992) has shown that the ornamental fish, tiger barb (*Barbus tetrazona*) and cherry barb (*B. titteya*) required carotenoids in their diet for sufficient colour development. Tsushima *et al.* (1998) observed that paprika was effective for pigmentation in goldfish and its effect was increased remarkably in the sunlight.

Tanaka *et al.* (1976) observed enhanced level of carotenoid in goldfish when fed with astaxanthin as a pigment source from crab waste. Hata and Hata (1976) have reported that in the Dutch lionhead goldfish, zeaxanthin was easily oxidized to astaxanthin, and carotenoid was also metabolized to astaxanthin.

Peimin *et al.* (1999) revealed that spirulina induced the growth and body colour of crucian carp. Phromkunthong (1988) reported that the addition of 10 per cent spirulina was highly effective in producing deeper colouration in fancy carp. Matsuno *et al.* (1980) suggested that the zeaxanthin might be metabolized to rhodoxanthin in vivo and the amount of astaxanthin in the integument of the test group whose diet contained zeaxanthin increased to eight times that of the control.

Ahilan and Jeyaseelan (2001) observed the colour enhancement in juvenile goldfish using shrimp head meal, dried drumstick leaves, dried curry leaves and turmeric powder. The carotenoid supplemented bitterlings had shown deep colouration (Kim *et al.*, 1999). Fey and Meyers (1980) reported the enhancement of colour in male gourami when fed with carotenoid as a pigment source.

Ahilan and Jegan (2008) observed colour enhancement in goldfish fed with vegitative additives *viz.*,coriander, mint and amaranth leaves.

Lovell (1992) has shown that the ornamental fishes, tiger barb (*Barbus tetrazone*) and cherry barb (*Barbus titteya*) require carotenoids in their diet for sufficient colour development to be marketable. He has shown that the carotenoids leutein and capaxanthin, from natural products which are used in commercial poultry feeds, can satisfactorily produce the commercially desirable yellow to deep red colours in these ornamental fish.

Research work at Central Institute of Freshwater Aquaculture (CIFA) identified four pigmented sources which enhances the fish colour. They are Coriander leaves, Curry leaves, Drumstick leaves, Marigold petal. These pigmented substances were added to the basal diets and fed to the young gold fish. Colouration was significantly better in marigold petal followed by other substances used in the experiment (Swain, 2005)

Artificial carotenoids source of five different types of leafy source of carotenoids from Agathi leaves, Curry leaves, Drumstick leaves, Amaranthus leaves and Colocasia leaves have been studied for enhancing rosy barb colouration (Naik and Swain, 2006).

Genetics and Pigmentation

In the world of ornamental fishes there are many vibrant colours attracting the public. Let's just take the colour red as an example and colour in koicarp. Skin pigmentation in koi is caused by dots (colour cells called chromatophores) the intensity of which is determined by how densely the dots are packed and how intensely each dot is coloured. A koi's genetic code will determine both factors, with the role of colour enhancement through feeding only being able to improve the colour of each colour cell (rather than increase their density in the skin). However, there is always hope as that same genetic code will also code for colour development where pattern and the appearance of other chromatophores may develop in the future. Koi carps are only able to exhibit pigments if they receive them (or their precursors) in their diet. Feed a koi a completely colour deficient diet, and over time, colours will fade to form a very 'unornamental' off-white skin colour. One more example is, when well pigmented gold fishes are kept in a clear water aquarium without any pigment diet, the colour start fading and after few days the fish looks very dull and unattractive. Carotenoids are the massive group of colour enhancing compounds that are stored and exhibited in fish skin. They are a group of chemicals that impart colour by the way they absorb and reflect light. Those that refract higher wavelengths of light (reds)are more desirable than those that refract the lower end of the spectrum (yellow). They are organic in nature, and are very closely related to Vitamin A, and similar in structure to vitamin E. Due to their similarity in chemical structure to these 2 vitamins, they behave in a similar way in living tissue, being very reactive and unstable, easily degraded in oxygen heat and light.

As carotenoids do degenerate over time, in the same way that a leaking bucket needs to be topped up to keep it full, koi require a constant supply of carotenoids to keep the chromatophores packed with carotenoids.

But unlike carp, they are unable to convert other pigments (such as lutein and carotene) into the desirable astaxanthin which must be fed directly in their diet. Koi absorb and convert carotenes very efficiently into astaxanthin, and is the secret behind the apparent anomaly of enhancing the red skin in koi by feeding them on green algae.

There are two approaches to enhancing colouration in koi - the scatter-gun approach (using natural sources of carotenoids) and the precise approach using synthetic colour enhancers.

Natural Sources

There are several recognised natural sources of carotenoids suitable for colour enhancement. Like any natural commodity, qualities and pigment content can vary from source to source, and being organic, can be liable to degradation during feed manufacture. However, natural sources are also renowned for offering a superb range of carotenoids giving koi (who have the ability to convert carotenoids) excellent colour potential. For example, marigold petals have more than 20 different carotenoids, which koi and goldfish can utilize properly. They also have a high concentration of these compounds (approximately 9000 mg per kilo), whereas shrimp or krill meal will only have about 200 mg per kilo, with the added issue of the exoskeletal material have an exceedingly high ash content.

Furthermore, there is a price to pay for koi and gold fishes using natural carotenoid sources as the process of converting them into more desirable astaxanthin requires energy.

Excess Feeding of Carotenoids

There is always a risk of overdosing of carotenoids in ornamental fishes. There has not much work been done on this aspect except on Koi carp. Few reports say that when koi carp were fed with carotenoids in excessive amounts, area of white tissue can turn muddy brown or yellow. This is caused by the effects carotenoids have of light. While white areas of skin will not deposit red pigmentation because they do not possess red chromatophores, high levels of carotenoids circulating in the tissue will inevitably interact with the light and cause whites to suffer. Premium koi carp diets will contain 'safe' levels of colour enhancers, giving the benefits of a precise and effective formulation.

Ingredients that Enhance the Koi Carp Colour

1. **Spirulina:** This has an above average carotenoid content that is easily assimilated, being found in very simple algae cells
2. **Paprika:** Red Pepper Meal. More potent than spirulina, and contains red pigments ready to be absorbed and assimilated immediately.
3. **Marigold Flower Meal:** Probably one of the most potent natural sources of carotenoids available. The pigments require manipulation by koi to convert them to reds.
4. **Artificial Colour Enhancers:** Astaxanthin and Canthaxanthin are guaranteed, potent sources of colour in the form that will be exhibited immediately in the skin.
5. **Yeast :** *Phaffia rhodozyma* This is easily digested and is a recognised source of carotene and astaxanthin.

Water quality may also play a supportive role in determining the colour of ornamental fishes. Stress causing by degraded water quality may dull the fish colours. A good quality biological filter and timely water exchange (75 per cent water replenishment at least once in a month depending upon the stocking size) will provide a better environment for displaying their brightest colours.

Conclusion

Indian ornamental fish trade is increasing steadily in the global market, among which the freshwater ornamental fish industry contributes only 0.008 per cent of the global trade. There are several obstacles contributes to the low production and low trade. Among various obstacles the ornamental fish industry in India facing today includes low seed production, less survival due to disease outbreaks, non availability of quality brood stock and commercial pigment enriched balanced feed for enhancing colours. Among the above parameters, colouration is an important factor, which plays a vital role in its commercialization of such beautiful fishes.

Chapter 9

Feed Additives for Colour Enhancement in Ornamental Fishes

B. Ahilan

Department of Aquaculture
Fisheries College and Research Institute, Thoothukudi – 628 008

Introduction

Ornamental fish production is an important component of the aquaculture industry. The FAO statistics indicated that the world export of ornamental fish steadily rose from US $ 160.7 million since 1999 to a peak of US $ 282.6 million in 2006. Most of the world's supplies of ornamental fishes are from Asian countries. Singapore is the largest exporter of ornamental fish contributing 21.70 per cent followed by Spain, Czech Rep and Malaysia during 2006 (Dey, 2008). USA is the world's largest single market for ornamental fishes and imported US $ 48.40 million worth of fish in 2006, followed by UK (US $ 30.80 million) and Japan (US $ 27.20 million). Ornamental fish production is the fourth largest sector in USA after catfish, trout and salmon.

The ornamental fish trade is a foreign exchange earner, besides being a source of employment. It has a significant role in the economy of developed and developing countries. The entire ornamental fish industry including accessories and feed are estimated to be worth of more than 14 billion US $ (Thomas, 2008).

Ornamental fishes are the world's most popular pets and fish keeping happens to be a popular hobby next only to photography. The fantastic shapes and brilliant colours of ornamental fishes won the heart of millions of people; hence, they can be aptly called as "Living Jewels". Most of the ornamental fishes are sourced from the developing countries in the tropical and subtropical regions, which contributed 60 – 65 per cent of the supply. With more than 120 countries involved in the ornamental fish trade and there are about 1,800 species of ornamental fish in the market of which over 1000 varieties are from freshwater origin, 90 per cent of the fresh water fishes are farmed while 10 per cent are collected from wild.

India is a tropical country and its contribution to global ornamental fish trade is negligible (Rs.5.7 crores, 0.5 per cent of international trade in 2005-06). A large number of freshwater ornamental fishes belong to family Cyprinidae from India are known to the world. Eight hundred and six species inhabit in freshwaters of India, out of which 196 species are considered as ornamental. At present, more than 50 per cent of ornamental fishes are exported from Northeastern region of India.

India is gifted with rich resources of water bodies and have 1, 64,153 km of rivers and canals, 19 million ha reservoirs, 2.0 million ha tanks and ponds, 1.5 million ha beels and oxbow lakes and 1.4 million ha brackish water. Also, it is bestowed with ideal climatic condition conducive for better growth, maturation and breeding of ornamental fishes.

The freshwater species dominating in the market are mainly from the families of Poecilidae, Characidae, Cyprinidae and Cichlidae. The trade in freshwater ornamental fish tends to grow more rapidly with special emphasis on coldwater fishes such as goldfish and koi carp.

Ornamental fishes are found to occupy a remarkable position in the international trade. The cost of ornamental fishes depends on its colour and appearance. Brightly coloured and attractive fishes always fetch high hence in the market.

Colouration in Fishes

Attractive colouration is a determining factor for the commercial value of ornamental fish. Pigments in the skin are responsible for coloration of fish. Carotenoids are the primary source of pigmentation in the skin of fishes. The yellow, red and blue colours are caused by various carotenoids or carotenoid protein complexes. But the fish are unable to synthesize carotenoids though they are able to modify them. So fish must get the carotenoids from their diet.

Fish in a natural environment meet their carotenoid requirements by ingesting aquatic plants (which synthesize carotenoids) or through the aquatic feed chain. Many carotenoids have been identified in the aquatic environment. The most usually common are lutein and zeaxanthin (yellow and deep orange) found in aquatic plants and plants and animals and astaxanthin (red) found in crustaceans. Fish that are grown or held in confinement where natural feed is limited or absent and artificial feed is the only source they may not have optimum colour.

Ornamental fish are usually grown in closed culture systems. Management convenience is the claimed advantage of this practice. The fish must be fed on commercial feeds. Feeds made from traditional ingredients are naturally or nutritionally sufficient but unless a satisfactory pigment source is added the fish will grow well but will lack adequate colouration.

Colouration is controlled by the endocrine and nervous system, but dietary sources of pigments also play a role in determining colour in fishes. The endocrine and nervous system both influence colouration in fish. The pituitary gland secretes hormones that direct the production and storage of pigments throughout the life of a fish and particularly as maturity is reached.

Pigment production and storage often increases at the onset of maturity. Many species use colour to provide amylase and attract mate. Fish of the family cichlidae are particularly. The automatic nervous system directs rapid colour changes in response to stimuli such as a predator or an aggressive tank mate.

Purpose of Colour

The colour, a fish display is multipurposeful *i.e.* it can be involved in :

1. Communication
2. Identification
3. Camouflage
4. Defence
5. Mimicry

Colour Enhancer

Colour enhancers are a natural pigments which when fed in the diet enhane the colour of fish flesh or skin. Carotenoids are the colour pigments responsible for enhancing skin colour some of which include alpha carotene, beta carotene, astaxanthin and canthaxanthin. Colour enhancers are used widely in the salmon and trout industry where they are used to enhance the pink in flesh.

Specialized pigment containing cells called chromatophores are located beneath the scales. These cells are branched, permitting pigment granules to the near or away from the surface and aggregated or dispersed. These cells are the reason for the variable and sometimes rapid changes in fish colour. Additionally, colourless purine crystals are contained in specialized chromatophores called iridophores. The iridophores are responsible for the silver green, particularly of small pelagic fish.

Carotenoid Pigments

Carotenoids and carotenoid – containing extracts have been widely used as colourants in feed for centuries. The colouration may be induced by direct addition to the feedstuff or indirectly by feeding to animals. Nowadays, the caroteniod containing plant materials are being used to enhance the colouration in feed fishes as well as aquarium fishes. Carotenoids are a family of over 600 natural lipid-soluble pigments that are produced in the micro algae, phytoplankton and higher plants. They also produce compounds such as essential fatty acid steroids, vitamin A, D, E and K (Goodwin, 1984; Davis, 1985). Some fish species such as Koi and various crustaceans have the enzymatic mechanisms to convert carotenoids into other forms as well such as astaxanthin – some fish/animals don't. Astaxanthin is the optional carotenoids for the proper pigmentation of the red/pink colours in aquaculture. Crustacean and other aquatic animals are unable to produce astaxanthin *de novo* naturally (Britton *et al.*, 1995).

Pigments are characterized by their colours. Carotenoid pigments are red and orange. Xanthophylls are yellow. Melanin pigments are black and brown. Phycocyanin is the blue pigment derived from blue-green algae. Cells containing yellow pigments overlying these containing blue pigments can produce green hues. Fish are capable of producing some pigments, but others must be supplied in the diet. Black and brown pigments are produced in cells called melanocytes. Fish are incapable of producing carotenoid and xanthophylls pigments. Therefore, these must be supplied in the diet.

Source and Types of Carotenoid Pigments

Natural sources of pigments are available in the diets of most fish. Colour enhancing diets may contain additional natural pigments to enhance colours of ornamental fishes. The types of carotenoid pigments include Betacarotene, Lutein, Taraxanthin, Astaxanthin, Tunaxanthin, Zeaxanthin and Alpha-beta doradexanthin.

The major sources of carotenoid pigments are *Chlorella vulgaris, Haematococcus plurialis, Spirrlina* sp., Shrimp head meal, marigold, paprika, carrot, pumpkin, capsicum, wheat, corn, alfalfa, krill extract, cray fish and red yeast (*Phaffia rhodozyma*).

The carotenoid pigments found in most marine and a few freshwater invertebrates is astaxanthin. This pigment gives the characteristic colour to the flesh of salmon and is available in the diet of aquarium shrimp and krill meals and salmon (fish) meal used as sourcer of protein in some feeds. Pure astaxanthin or canthaxanthin (synthetic astaxanthin) may also be added to fish feed to enhance red and orange colouration. These carotenoid pigments are often added to feeds for farm raised salmon and trout to give fillets a desirable red colour.

Xanthophylls (yellow pigments) are found in corn gluten meal and dried egg that may be added to the diet to enhance yellows. The ground petals of marigold flowers have also been used as a source of xanthophylls. The blue-green algae spirulina is a such source of phycocyanin and may be added to a diet to enhance blue colouration. The expense of supplementary pigments often limits the amount used in trophical fish feeds.

Testosterone also plays a role in enhancing the colour of fishes. Testosterone supplied in the diet likely allows a premature storage and expression of pigments in the chromatophores. Fish that offers exhibit drab juvenile colouration may then show full adult colouration.

Carotenoids in Plants

The natural extracts first used were those of carrot, palm oil saffron, tomato and paparika (Britton, 1996). The powdered, dried plant materials and extracts of them have been used for many years. Generally green leaves contain large amount of carotenoid. Carotenoids are often considered to be plant pigments alone but also occur widely in bacteria, fungi, algae and in animals. Goodwin (1980) provided extensive details about the distribution about carotenoids in plants and animals.

In plants, the carotenoids occur universally in the chloroplast of green tissues, but their colour is masked by the chlorophylls. The leaves of virtually cell species contain the same main carotenoids, that is β-carotene (usually 25-30 per cent of the total), lutein (around 45 per cent), violaxanthin (15 per cent) and neoxanthin (15 per cent) (Britton,1996). Small amount of α-carotene, α-β-cryptoxanthin, zeaxanthin, antheraxanthin and lutein 5, 6 epoxide are also frequently present (Cogdwell, 1988). Carotenoids are also widely distributed in non-photosynthetic tissues of plants and are responsible for the yellow, orange and red colour of many flowers and fruit. They are usually located in chromoplasts (Goodwin, 1980). Natural extracts of some flower such as marigold (*Tagetes erecta*) contain large amount of lutein and are used for colouration.

Total carotenoid content varied from species to species. Chandersekar *et al.* (1999) reported, the carotenoid content in amaranthus, coriander and mint and the levels were found to be 37,400 μg/100 g, 15,077 μg/100 g and 8,305 μg/g respectively in raw form. They used spectrophotometry and high performance liquid chromatography for the analysis. Kowsalya *et al.* (2001) reported that the total carotenoid content of these were 29.064 mg/100 g, 36.720 mg/100 g and 29.057 mg/100 g respectively on the basis of dry weight (sun dried).

Amaranths are hardy, wild, uncultivated, fast growing cereal – like plants. The growth habits vary from horizontal to erect and branched to unbranched; their leaves and stems are of many shades of green and seeds was from black to white. Amaranth is one of the few double duty plants which can supply tasty leafy vegetables, as well as grains of high nutritional quality. The leaves of all 50-60 species of amaranths are edible (Singhal and Kulkarni, 1988).

Pigmentation on Feed Fishes

The sole feeding of a formulated feed which contains 2 per cent of an oil extracted from the meal of the antartic krill *Euphausia superba* to yellow tail *Seriola quinqueradiata* induced poor dark green colour at the back and sides unlike the natural one of iridescent blue-green (Fujita *et al.*, 1983).

Inoue *et al.* (1988) observed that the oil which is extracted from cray fish carapace with soy oil, enhanced pink colour pigment deposition in rainbow trout *Salmo garidneri*. The pigmentation of cultured red sea bream *Pagrus major* with the antartic krill is known to be attained by feeding any of the frozen krill, krill meal and oil extracted from meal (Fujita *et al.*, 1983).

Kamata *et al.* (1990) used the flower (*Adonis aestivalis*) and its extract as pigment source for rainbow trout at the level of 100 mg pigment/100 g diet. The pigmentation of the flesh of cultured coho salmon *Oncorhynchus kisutchi* was examined on a practical scale by feeding diets supplemented mainly with antartic krill and littoral mysid both of which are rich in a red carotenoid astaxanthin (Mori *et al.*, 1990).

Pigmentation of cultured sweet smelt fed diets supplemented with a blue green algae *Spirulina maxima* was carried-out by Mori *et al.* (1987). He reported that spirulina enhanced the pigment deposition in sweet smelt. Tveranger (1986) observed the effect of pigment content in brood stock diet on subsequent fertilization rate, survival and growth rate of rainbow trout offspring by using 10 per cent krill meal as a source of pigmentation.

The effect of supplementing trout diet with astaxanthin rich micro algae *Haematococcus pluvialis* was studied by Sommer *et al.* (1991). He concluded that the total carotenoid levels were significantly higher.

Metusalach *et al.* (1996) carried out a study on deposition and metabolism of dietary canthoxanthin in different organs of arctic charr (*Salvelinus alpinus* L). The result showed that fish flesh was the major tissue for storing carotenoids followed by skin, liver and gonad. Nickell and Bromage (1998) studied the efficiency of pigmentation, the variation and development of fillet colour in rainbow trout fed with astaxanthin enriched diets at different stages of fry development and for different periods of time.

The effect of water temperature on pigmentation of rainbow trout was investigated by feeding a test diet with astaxanthin. The carotenoid content of fish was more at 15^{0}C than at 5^{0}C but no significant temperature effects on the carotenoid concentration in the skin, liver and gut were observed by Kyeon and Storebakken (1991). Flesh pigmentation was measured against a colour fan in chinook salmon (*Oncorhychus tshacoytscha*) by using carotenoid as a pigment source (McCallum *et al.*, 1987).

The retention of carotenoids in the rainbow trout was significantly different for the two carotenoid sources (retention of astaxanthin 11.4 per cent, canthaxanthin 7.1 per cent) (Storebakken and Chouhert, 1991).

Torrisen (1986) reported that astaxanthin was observed more in the flesh than canthaxanthin. The trout fed with canthaxanthin tended to accumulate more carotenoids in the skin as metabolites (Bjerkeng *et al.*, 1990). The wet diet resulted in a lower pigmentation than the dry diet in salmon (Storebakken *et al.*, 1986).

Skin of fish fed with astaxanthin mainly contained astaxanthin esters and skin of fish fed with canthaxanthin contained reductive metabolites of canthaxanthin (Bjerkeng *et al.*, 1992). Marron or fresh water cray fish *Cherax tenuimanus* fed on the algal slurry (Microalgae, *Dunaliella salina*) had elevated levels of total carotenoid and β-carotene level (Sommer *et al.*, 1991). Boonyaratpalin and Unprasest (1989) tested four pigments such as spirulina, marigold petal meal, shrimp head meal and turmeric on red tilapia, *Oryochromis niloticus*. The results showed single red, golden flame, flame and orange colouration respectively.

Choubert Jr. (1979) reported that the incorporation of spirulina algae in rainbow trout causes a yellow – trout pigmentation. The retention co-efficient decreased as the pigment dose in the diet increased (Choubert *et al.*, 1989). Ibraham *et al.* (1984) investigated the pigment enhancement in cultured red sea bream using antartic krill. Oil extraction from antartic krill was used as colour enhancing ingredients in diet (Arai *et al.*, 1987).

Pigmentation on Ornamental Fishes

Colouration is the determining factor in ornamental trade. Natu Rose (*Haematococcus pluvialis*) is a safe natural source of astaxanthin derived from a unique strain of the microalgae. It has been extensively tested and has demonstrated exceptional pigmentation in Koi and other tropical fishes (Ronneberg *et al.*, 1979). Marine ornamental fish obtains carotenoids by feeding upon algae, coral or prey that has accumulated these pigments (Fox.D. Biochromy, 1979). The red varieties of goldfish, stores significant quantities of astaxanthin in the skin, which can also be increased by inclusion of an artificial diet (Fox.D. Biochromy, 1979).

Ako *et al.* (1999) observed a colour enhancement in fresh water red velvet swordtail (*Xiphophorus helleri*), rainbow fish (*Pseudomugil turcatus*) and Topaz cichlids (*Cichlasoma myrnae*) by using diets containing *spirulina platensis* (1.5 – 2 per cent) and 1 per cent *Haematococus pluvialis*.

Tanaka *et al.* (1976) observed enhanced level of carotenoid in goldfish when fed with astaxanthin as a pigment source from crab waste. Hata and Hata (1976) have reported that in the Dutch lionhead goldfish, zeaxanthin was easily oxidized to astaxanthin, and b-carotenoid was also metabolized to astaxanthin.

Bitzer (1963) found out that paprika not only enhanced the colour of the fish, but also increased hatchability. The basic synthesis of carotenoids is carried-out in algae which are transferred to fish via. crustaceans and insects (Brinchmann, 1967).

Sherief and Mathew (1996) observed colour enhancement in goldfish using different ingredients such as carrot meal, beetroot meal and red grape skin meal. Lovell (1992) has shown that the ornamental fish, tiger barb (*Barbus tetrazona*) and cherry barb (*B. titteya*) required carotenoids in their diet for sufficient colour development. Tsushima *et al.* (1998) observed that paprika was effective for pigmentation in goldfish and its effect was increased remarkably in the sunlight.

Peimin *et al.* (1999) revealed that spirulina induced the growth and body colour of crucian carp. Phromkunthong (1988) reported that the addition of 10 per cent spirulina was highly effective in producing deeper colouration in fancy carp. Matsuno *et al.* (1980) suggested that the zeaxanthin might be metabolized to rhodoxanthin *in vivo* and the amount of astaxanthin in the integument of the test group whose diet contained zeaxanthin increased to eight times that of the control.

Ahilan and Jeyaseelan (2001) observed the colour enhancement in juvenile goldfish using shrimp head meal, dried drumstick leaves, dried curry leaves and turmeric powder. The carotenoid supplemented bitterlings had shown deep colouration (Kim *et al.*, 1999). Fey and Meyers (1980) reported the enhancement of colour in male gourami when fed with carotenoid as a pigment source.

Ahilan and Jegan (2008) observed colour enhancement in goldfish fed with botanical additives *viz.*,coriander, mint and amaranth leaves.

P.M.Sherief and Mathew, they have done a research on gold fish by using different colour ingredients mixed with feed. They reported that, carotene mixed feed (carrot meal) shows higher colour development with reference to carotenoid content compared to other sources (beet root meal as a source of betalaine and red grape skin as a source of anthocyanin).

Lovell (1992) has shown that the ornamental fishes, tiger barb (*Barbus tetrazone*) and cherry barb (*Barbus titteya*) require carotenoids in their diet for sufficient colour development to be marketable. He has shown that the carotenoids leutein and capaxanthin, from natural products which are used in commercial poultry feeds, can satisfactorily produce the commercially desirable yellow to deep red colours in these ornamental fish.

Colour Enhancement of Ornamental Fish

The colour of ornamental fish may be enhanced by following methods.

1. Dyeing
2. Painting
3. Feeds with hormone
4. Feeds with pigments

Dyeing

Dyeing colorless fish has recently become popular. The fish are immersed in water containing dye and the immersion and handling may lead to disease problem.

Painting

The practice of painting essentially colourless fish has become widespread. The neon coloured paint is non-toxic but the handling and painting, coupled with shipping stress often invites disease problems. These fish often contract ich (Icthyophthirius multifilis) and fungal infection. The paint is shed in times and the fish returns to being colourless which may be more disturbing to some one paying a premium for painted fish.

Feeds with Hormone

Hormones may be used to enhance fish colouration by causing a false early maturity. Testosterone supplied in the diet likely allows a premature storage and expression of pigments in the chromatophores. Fish that often exhibit drap juvenile colouration may then show full adult colouration. Fish treated with hormones often become all male, sterile and require a continuous dietary supply of hormones to maintain coloration.

Feeds with Pigments

Carotinoids are the group of over 600 natural lipid soluable pigments that are primarily produced with in phytoplankton, algae and plants. This pigments are responsible for the broad variety of colours in nature. Most notable are brilliant yellow, orange and red colours of fruits, leaves and aquatic animals. Among all of the numerous classes of natural colours, the carotinoids are the most wide spread and structurally divers pigmenting agents. Although plants algae and some fungal and bacterial species synthesize carotenoids, animals can not produce them de novo.

Spirulina, marigold meal, alfalfa, krill and other curious products are often included in the diet as colour enhancers. Synthetic colour enhancers provide a guaranteed content of specific carotinoids but do not offer the wide range of specific carotinoids found in the natural products.

Conclusion

The use of carotinoids as pigments in aquaculture is well documented, and it appears their broader functions include a role as an antioxidant, provitamin A activity, immune enhancement, hormones, reproduction, growth, maturation and photoprotection. Carotenoids cannot introduce new colours in lipophores which are not coded for genetically *i.e.* an orange fish can not be made red by feeding excessive quantity of colour enhancers. In conclusion, good skin quality can only be achieved by feeding a high quality balanced diet while maintaining an optimum water quality environment. A good quality colour enhancing diet will contain a wide range of quality colour enhancer and improve the colouration on a healthy fish while not adversely affecting white areas. Water quality may also play a support role in determining the colour of ornamental fish. Degraded water quality increase stress on capture fish and may dull fish colours. A high quality biological filter and routine at least be weekly water changes will peroside an enshrinements enabling fish displaying their highest colours.

Feeding a varied diet rich in sources of pigments along with good water quality will thus ensure the fish to develop good colours.

References

Ahilan, B. and Prince Jeyaseelan, M.J. 2001. Effects of different pigments sources on colour changes and growth of juvenile *Carassius auratus*. *Jour. Aqua. Trop.* 16(1): 29-36.

Ako, H., Asano, L., Fukada, M. and Tamaru, C.S. 1999. Culture of the rainbow fish *Pseudomugil furcatis* and the use of a Hawaii-specific feed. *Tropical Gold, Ia O Hawaii.* 2: 1-9.

Arai, S., Mori, T., Miki, W., Yamaguchi, K., Konosu, S., Satake, M. and Fujita, T. 1987. Pigmentation of juvenile coho salmon with carotenoid oil extracted from antartic krill. *Aquaculture*. 66: 255-264.

Bjerkeng, B., Storebackken, T. and Liaaen-Jensen, S. 1990. Dose response to carotenoids by rainbow trout in the sea: resorption and metabolism of dietary astaxanthin and canthaxanthin. *Aquaculture*. 91: 153-162.

Bjerkeng, B., Storebakken, T. and Jensen, L.S. 1992. Pigmentation of rainbow trout from start feeding to sexual maturation. *Aquaculture*. 108: 333-346.

Boonyaratpalin, M. and Unprasert, N. 1989. Effect of pigments from different sources on colour changes and growth of red *Oreochromis niloticus*. *Aquaculture*. 79: 375-380.

Boonyaratpalin, M., Thongraod, S., Supamattaya, K., Britton, G. and Schlipalius, I.F. 2001. Effects of b-carotene source, *Dunaliella salina* and astaxanthin on pigmentation, growth, survival and health of *Penaeus monodon*. *Aquaculture Research*. 32: 182-190.

Britton, G. 1996. Carotenoids. In: Hendry and Houghton (eds.) Natural food colorants. Blackie Academic and Professional. pp: 197-240.

Britton, G., Jensan, L.S. and Pfander, H. 1995. Carotenoids today and challenges for the future. In. Britton, G., Jensan, L. S. and Pfander, H. (eds.), Carotenoids Vol. I A Isolation and Analysis. Basal: Birkhauser.

Choubert, G. and Storebakken, T. 1989. Dose response to astaxanthin and canthaxanthin pigmentation of rainbow trout fed various dietary carotenoid concentrations. *Aquaculture*. 81: 69-77.

Choubert, G. Jr. 1979. Tentative utilization of spirulina algae as a source of carotenoid pigments for rainbow trout. *Aquaculture*. 18: 135-143.

Cogdwell, R.J. 1988. In: Goodwin, T.W. (ed.), Plant pigments, Academic Press, London. pp: 183-230.

Fey, M. and Meyers, S.P. 1980. Evaluation of carotenoid fortified flake diets with the pearl gourami, *Trichogaster leeri*. *Journal of Aquaculture*. 1: 15-19.

Fujii, R. 1969. Chromatophores and pigments. In: Hoar, W.S. and Randal, D.J. (Eds), Fish Physiology, Vol. 3, Academic Press, New York. pp: 307-353.

Fujita, T., Fatake, M., Hikichi, S., Takedu, M., Shimeno, S., Kuwabara Miki, H.W., Yamaguchi, K. and Konosu, S. 1983. Pigmentation of cultured yellow tail with krill oil. *Bull. Jap. Soc. Sci. Fish.* 49: 1595-1600.

Fujita, T., Fatake, M., Watanabe, T., Kitajima, T., Miki, W., Yamaguchi, K. and Konosu, S. 1983. Pigmentation of cultured red sea bream with astaxanthin diester purified from krill oil. *Bull. Jap. Soc. Sci. Fish.* 49: 1855-1861.

Goodwin, T.W. 1951. Carotenoids in fish. In: The biochemistry of fish. *Bio chem. Soc. Symp.* 6: 63-82.

Goodwin, T.W. 1980. The biochemistry of the carotenoids. Vol. 1, Plants, Chapman and Hall, London.

Goodwin, T.W. 1984. The biochemistry of the carotenoids, 2nd ed., Chapman and Hall, London. pp: 64-96.

Hata, M. and Hata, M. 1976. Carotenoid metabolism in fancy red carp. *Cyprinus carpio* – II. Metabolism of C - Zeaxanthin. *Bull. Jap. Soc. Sci. Fish.* 42: 203-205.

Inoue, T., Simpson, K.L., Tanaka, Y. and Sameshima, M. 1988. Condensed astaxanthin of pigmented oil from crayfish carapace and its feeding experiment. *Nippon Suisan Gakkaishi.* 54: 103-106.

Iwahashi, M. and Wakui, H. 1976. Intensification of colour of fancy carp with diet. *Bull. Jap. Soc. Sci. Fish.* 42: 1339-1344.

Kamata, T., Neamtu, G., Tanaka, Y., Sameshima, M. and Simpson, K. 1990. Utilization of *Adonis aestivalis* as a dietary pigment source for rainbow trout, *Salmo gairdneri*. *Nippon Suisan Gakkaishi*. 56: 783-788.

Kim, H.S., Kim, Y.H., Cho, S.H. and Jo, J.J. 1999. Effects of dietary carotenoids on the natural colour of the bitterling (*Rhodeus uyekii*), *J. Korean fish. Soc.* 32: 276-279.

Kowsalya, S., Chandrasekhar, U. and Balasasirekha, R. 2001. Beta carotene retention in selected green leafy vegetables subjected to dehydration. *Ind. J. Nutr. dietet*. 38: 374-383.

Lonneberg, E. 1929. Einige studein uber die lipochrome der fishche, *Ark Zool. Bd.* 6: 1-10.

Lovell, T. 1992. Dietary enhancement of colour in ornamental fish. *Aquaculture Magazine.* 18: 77-79.

Matsuno, T., Katsuyama, M., Iwahashi, M., Koike, T. and Okada, M. 1980. Intensification of color of red tilapia with lutein, rhodoxanthin and spirulina. *Bull. Jap. Soc. Fish.* 46(4): 479-482.

McCallum, I.M., Cheng, K.M. and March, B.E. 1987. Carotenoid pigmentation in two strains of chinook salmon (*Oncorhynchus tshacoytscha*) and their crosses. *Aquaculture.* 67: 291-300.

Metusalach, Synowiecki, J., Brown, J. and Shahidi, F. 1996. Deposition and metabolism of dietary canthaxanthin in different organs of arctic charr. (*Salvelinus alpinus* L). *Aquaculture.* 142: 99-106.

Mori, T., Muranaka, T., Miki, W., Yamaguchi, K., Konesu, S. and Watanabe, T. 1987. Pigmentation of cultured sweet smelt fed diets supplemented with a blue green alga *Spirulina maxima. Nippon Suisan Gakkaishi.* 53: 433-438.

Olsen, J.A. 1979. A simple dual assay for vit. A and carotenoids in human and liver. *Nutrition Report Journal.* 19: 807-813.

Parker, G.H. 1948. Animal colour changes and their neurohumours. Cambridge Univ. Press. London, New York.

Peimin, H., Yinjiong, Z. and Wenhui, H. 1999. Effect of the spirulina feed on the growth and body colour of crucian carp. *J. Fish. China/Shuichan Xuebao.* 23: 162-168.

Peterson, W.J., Hughes, J.S. and Payne, L.F. 1939. The carotenoid pigment. Agriculture experimental station, Kanas State College of Agriculture and Applied Science. *Technical Bulletin.* 46: 5.

Phromkunthong, W. 1988. Effect of carotenoid pigments from different sources on colour changes of fancy carp, *Cyprinus carpio.* Abstracts of Master of Science Thesis (Fisheries Science), Fac. Fish. Kasetsart Univ. Bangkok. P: 1.

Sherief, P.M. and Mathew, P.M. 1996. Dietary enhancement of colour in goldfish (*Carassius auratus*). *Fishing Chimes.* 16: 17-19.

Singh, T. and Dey, V.K. 2003. Ornamental fish trade runs into billions. *Info fish Int.*, 5: 54-60.

Singhal, R.S. and Kulkarni, P.R. 1988. Review: Amaranths – an underutilized resource. *Int. J. Food Sci. Tech.* 23: 125-139.

Sommer, T.R., Morrissy, N.M. and Potts, W.T. 1991. Growth and pigmentation of marron (*Cherax tenuimanus*) fed a reference diet supplemented with the microalgae, *Dunialiella salina. Aquaculture.* 99: 285-295.

Sommer, T.R., Potts, W.T. and Morrissy, N.M. 1991. Utilization of microalgal astaxanthin by rainbow trout (*Oncorhynchus mykiss*). *Aquaculture.* 94: 79-88.

Sommer, T.R., Souza, F.M. L.D. and Morrisssy, N.M. 1992. Pigmentation of adult rainbow trout, *Oncorhynchus mykiss* using the green alga, *Haematococcus Pluvialis*. *Aquaculture*. 106: 63-74.

Steven, D.M. 1948. Studies on animal carotenoids I. Carotenoids of the brown trout (*Salmo trutta* Linn.) *J. Exp. Biol.* 25: 369.

Storebakken, T. and Choubert, G. 1991. Flesh pigmentation of rainbow trout fed astaxanthin or canthoxanthin at different feeding rates in freshwater and seawater. *Aquaculute*. 95: 289-295.

Storebakken, T., Foss, P., Huse, I., Wandsvik, A. and Lea, T.B. 1986. Carotenoid in diets for salmonids. III. Utilization of canthaxanthin from dry and wet diets by Atlantic salmon, rainbow trout and sea trout. *Aquaculture*. 151: 245-255.

Thongrod, S., Tansutapanich, A. and Torrissen, O.J. 1995. Effects of dietary astaxanthin supplementation on accumulation, survival and growth in postlarvae of *Penaeus monodon* fabricus, Larvi 95 – Fish and shellfish larviculture symposium, Gent. Belgium. *Eur. Aqua. Socie. Special. Bull.* pp: 24.

Torrissen, O.J. 1986. Pigmentation of salmonids – A comparison of astaxanthin and canthaxanthin as pigment source for rainbow trout. *Aquaculture*. 53: 271-278.

Tsushima, Miyuki, Nemato, Hidetada, Matsuno. and Takao. 1998. The accumulation of pigments from paprika in the integument of goldfish *Carrasius auratus*. *Fish. Sci.* 64: 656-657.

Tswett, M. 1911. Chlorophyll in plants and in the animal kingdom, Warschene. 8: 379.

Tveranger, B. 1986. Effect of pigment content in broodstock diet subsequent fertilization rate, survival and growth rate of rainbow trout (*Salmo gairdneri*) offspring. *Aquaculture*. 53: 85-93.

Willstatter, R. and Mieg, W. 1907. A method for the separation and determination of chlorophyll derivatives. Ann. 350: 1-47.

Yamada, S., Tanaka, Y., Sameshima, M. and Ito, Y. 1990. Pigmentation of prawn (*Penaeus japonicus*) with carotenoids; I Effect of dietary astaxanthin, β-carotene and canthaxanthin on pigmentation. *Aquaculture*. 87: 323-330.

Part VI

Fish Nutrition in Aquariculture

Chapter 10

Live Feeds in Ornamental Fish Farming

S. Felix

Fisheries Research and Extension Centre,
Tamil Nadu Fisheries University, Chennai – 51

There are different types of aquarium Live Feed that can be used to supplement the diets of ornamental fishes. There are many benefits of using live feed:

- ☆ Providing well-balanced and extra nutritional requirement
- ☆ Rich source of carotenoids
- ☆ High nutritional value and protein content.

Live fish feeds include earthworms, sludge worms, water fleas, bloodworms, white worms and brine shrimp. Feed for larvae and young fish fry include infusoria (Protozoa and other microorganisms), newly hatched brine shrimp and micro worms. These are the most preferred type of feed for fishes, but are difficult to get. However, freeze dried forms of earthworms, tubifex and bloodworms are now available in the market.

Culture of Live Feeds

1. Infusoria

Infusorians are minuscule single cell organisms that live in water. To cultivate Infusoria we only need tap water, some vegetables and a few clean containers such as jars are required.

Figure 10.1: Infusoria Culture.

Boil the tap water and let it cool down to room temperature. Fill the jars with the water and add some bruised vegetables to each jar. Cabbage leaves or banana skins are recommended, since they are known to produce a lot of infusoria. One banana skin or 3-4 lettuce leaves is a good amount for each jar. Do not put any lids on the jars. Place your jars in a warm place where they will receive moderate light.

The water in the jars will soon turn cloudy and begin to smell. As the infusorians develop the smell will become increasingly sweet and the water will clear up. It will usually take3-4 days for Infusoria to develop. When the water looks clear, you can start feeding your fry with the Infusoria. Use a siphon or a spoon to remove the Infusoria from the top of the jar and place it in your fry aquarium. Make sure not to bring any plant material from the infusoria jars over to the fry rearing tank, since this can pollute the water. Start developing new Infusoria culture every 3-4 day to make sure you always have enough feed for your fry until they are big enough to eat larger feed.

2. Brine Shrimps (*Artemia salina*)

Baby brine shrimp is a very important feed for most fish fry. Most livebearers will eat a lot of baby brine shrimp, and this definitely makes them grow faster and healthier even if livebearers are among the fry that can be raised on powdered flake feed alone. The eggs of brine shrimp are carefully processed and collected and are sold in many pet stores. The aquarist can purchase these eggs, hatch them and use them to feed fry. One advantage in hatching the eggs at home is that these can then be fed special additives that will directly be transferred to your fry. Some fry are too small to be able to eat newly hatched brine shrimp and such fry will need infusoria the first few days before they can start eating brine shrimp.

Most fish fry are large enough to eat brine shrimp as their first feed, especially if you give them newly hatched brine shrimp. Very small fry should instead be fed infusoria, and brine shrimp can be introduced as a second feed. If you feed your brine shrimp suitable feed, they will grow large enough to be used as feed for adult fish as well.

Hatching Brine Shrimp

Although using Artemia cysts appears to be simple, several factors are critical for hatching the large quantities needed in larval fish production. They include cyst disinfection or decapsulation prior to incubation, and hatching under the following optimal conditions constant temperature of 25–28°C, 15–35 ppt salinity, minimum pH of 8.0, near saturated oxygen levels, maximum cyst densities of 2 gm per l, and strong illumination of 2000 lx. All these factors will affect the hatching rate and maximum output, and hence, the production cost of the harvested Artemia nauplii. After hatching, and prior to feeding them to the larvae, Artemia nauplii should be separated from the hatching wastes. After switching off the aeration in the hatching tank, cyst shells will float and nauplii will concentrate at the bottom of the tank. They should be siphoned off within 5– 10 min and thoroughly rinsed with seawater or freshwater, preferentially using submerged filters to prevent physical damage to the nauplii. On a commercial scale, the separation of nauplii from cyst shells is performed with a standpipe perforated a few centimeters from the bottom. The free- swimming nauplii on top of the unhatched cyst are evacuated through the perforation, while the unhatched cysts are kept out of the turbulent area. The nauplii are further concentrated in a concentrator rinser and separated from the last cysts on a double screen. When decapsulated cysts are used, the membranes are generally skimmed off by the use of high performant airstones.

Figure 10.2: Hatching of Artemia Cysts.

Figure 10.3: Hatched Out Artemia Nauplii.

In their first stage of development, Artemia nauplii do not feed but consume their own energy reserves. At the high water temperatures that are applied during cyst incubation, freshly hatched *Artemia nauplii* develop into the second larval stage Instar II metanauplii. Within 6–8 h. it is important to use first-instar nauplii for feeding, rather than starved second-instar metanauplii, which are transparent and less visible. Instar II metanauplii are about 50 per cent larger in length and swim faster than first instars. As a result, they are less acceptable as prey. Furthermore, they contain lower amounts of free amino acids, so they are less digestible and with lower individual dry weights.

Storing freshly hatched nauplii at temperatures near 4°C, in densities of up to eight million nauplii per liter for up to 24 h will greatly reduce their metabolic rate, *i.e.*, only 2.5 per cent drop in individual dry weight versus 30 per cent at 25°C, and preclude molting to the second instar stage. This 24-h cold storage economizes the Artemia cyst hatching effort *e.g.*, fewer tanks, larger volumes, a maximum of one hatching and harvest per day and allows not only a constant supply of a high-quality product but also the possibility of more frequent feed distributions. This is beneficial for fish larvae because feed retention time in larviculture tanks can be reduced and hence the growth of Artemia in the culture tank minimized.

Enrichment of Artemia

Taking advantage of the primitive feeding characteristics of Artemia nauplii, it is possible to manipulate the nutritional value of HUFA-deficient Artemia. Since brine shrimp nauplii that have molted into the second instar stage *i.e.*, about 8 h following hatching are non-selective particle feeders, simple methods have been developed to incorporate different kinds of products into the Artemia prior to feeding to predator larvae. This method of bioencapsulation, also called Artemia enrichment or boosting, is widely applied in marine fish and crustacean hatcheries for enhancing the nutritional value of Artemia with essential fatty acids. Researchers developed enrichment products and procedures using selected microalgae, micro-encapsulated products, yeast, emulsified preparations, self-emulsifying concentrates, and micro-particulate products, either singly or in various combinations. The highest enrichment levels are obtained from emulsified concentrates freshly hatched nauplii are transferred to the enrichment tank at a density of 100– 300 nauplii per rml for enrichment periods for 24 h, respectively. The enrichment medium consists of hypochlorite-disinfected and neutralized seawater maintained at 25°C. The enrichment emulsion is added in consecutive doses of 0.3 gm for every 12 hours. Strong aeration using airstones or pure oxygen is required to maintain dissolved oxygen levels above 4 ppm. Enriched nauplii are harvested after 24 or 48 h, thoroughly rinsed and stored at temperatures below 10°C to assure that HUFAs are not metabolized during storage. Enrichment levels of 50–60 mgm n-3 HUFAs are obtained after 24-h enrichment with the emulsified concentrates. Nauplii should be transferred or exposed to the enrichment medium as soon as possible before first feeding, so they begin feeding immediately after the opening of the alimentary tract Zinstar II stage. As a result, the increase of nauplius size during enrichment can be minimized, *i.e.*, after 24-h enrichment, GSL *Artemia nauplii* will reach about 660 mm, and after 48-h enrichment, about 790 mm.

The nutritional quality of commercially available Artemia strains being relatively poor ineicosapentaenoic acid, EPA, 205n-3. and especially docosahexaenoic acid DHA, 226n-3. It isessential and common practice to enrich these live prey with emulsions of marine oils. In Artemia, the most commonly applied boosting technique is a 24-h enrichment period after hatching. However, that the enrichment technique has limitations as Artemia are selectively catabolizing some of the nutrients such as DHA and phospholipids. Nowadays, various enrichment emulsions have been formulated differing in the fatty acid composition of their

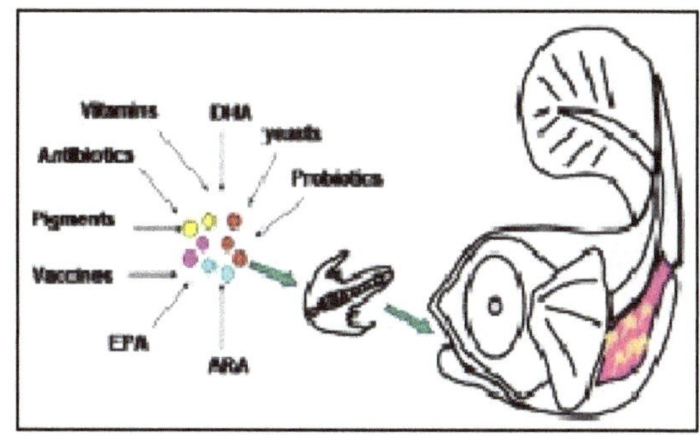

triglycerides. In this respect, the traditional formulations rich in EPA have been replaced by new products rich in DHA and arachidonic acid.

To reduce the risks for oxidation of these fatty acids, higher concentrations of vitamin E are incorporated into the emulsions. Also, vitamin C has been incorporated in booster formulations that increase the level of ascorbic acid in Artemia to 2000 ppm. All these changes in the formulation of the enrichment diets offer more possibilities to cover the needs of different species and help to reduce problems related to diseases, stress resistance, malformation, and pigmentation in numerous fish species. its palatability induces a good and fast feeding response. These characteristics, coupled with the use of bioencapsulation techniques to enhance the quality of the on-grown Artemia, make this organism an optimum diet for nursery of the fish.

Control on the Bacterial Input from Cysts and Nauplii

Chemical decapsulation of Artemia cysts using hypochlorite is a widely applied technique in hatcheries. Besides its advantages, for zootechnical reasons (easier separation) prevention of gut obstruction, direct use of decapsulated cysts., it is believed to have a beneficial effect on hatching and provide a complete disinfection of the cysts. The use of heterotrophically grown microalgae and its extracted oil as sources of nutrients and essential fatty acids for rotifers and Artemia, which were then used in fish larvae as well as in formulated broodstock diets. These sources can be particularly rich in docosahexaenoic acid (DHA) and arachidonic acid (ArA), which are required for larval growth and survival as well as contributing to egg and sperm quality when included in broodstock diets. Heterotrophically grown algal and fungal supplemented diets are highly effective in delivering essential fatty acids either through larval live feed enrichment or directly through the fish diet.

Microworms

Microworms are also another option to use as cheap aquarium live fish feed. They can be easily cultivated using packets of cooked oatmeal added with small amount of yeast. Usually a common method to grow microworms involves using plastic feed containers with the cover punched with holes for air circulation purpose. Add the starter culture together with the yeast plus your oatmeal mixture and then wait for two to three days for the growth to take place. On opening the container cover, the microworms can be seen crawling up the sides of the containers and some on the cover surface. Harvest the worms by scraping them off with a clean butter knife and then dip it into aquarium water. Microworms are well known to be an excellent feed source for first feeding fish larvae. The species most commonly cultured in the aquarium hobby is believed to be *Panagrellus redivivus*, a member of the nematode family, Panagrolaimidae. *Panagrellus redivivus* is a nematode which is easy to rear in large quantities in culture. The worms feed on bacteria which are precultured. They have a short life cycle and a high fecundity. *Panagrellus redivivus* are a tiny nematode about 0.5 to 2.0 mm in length

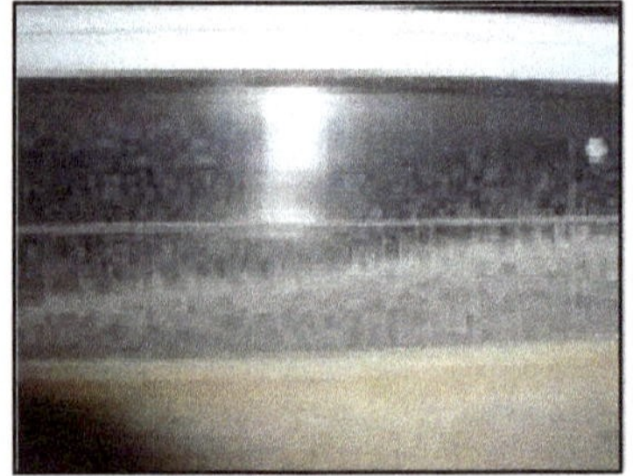

Figure 10.4: Microworms.

and 0.05 mm in diameter. They reproduce sexually and are livebearers; releasing 10to 40 young every 5 to 7days for a 26 to 36 day life span. The young reach sexually maturity in approximately three days.

Their size increases by three times during the first day and five to six times during the next three days. Microworms have been cultured by aquarists since the early 1930's as a live feed for a variety of fish species. Their small size and ease of culture has received renewed attention in recent years with rising costs and declining hatch rates of brine shrimp eggs sold in the aquarium industry. Microworm has a better nutritional profile to that of brineshrimp, containing 48 per cent protein, 21 per cent lipids, 7 per cent glycogen, 1 per cent organic acids, and 1 per cent nucleic acids. Approximately 70 per cent of the lipids are fatty acids and the remainder is phospholipids.

Culture of Microworms

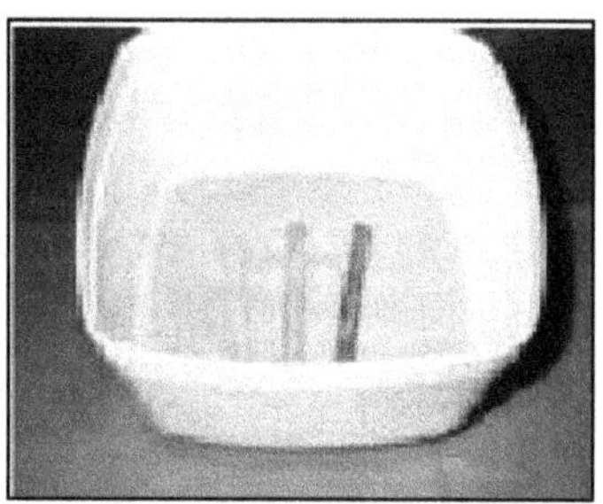

Figure 10.5: Culture of Microworms.

Bread soaked in water with yeast and oats can be used as feed. Microworms can be cultured in almost any shallow, flat, watertight container with a tight fitting lid. This prevents contamination by insects and also prevents the culture from dehydrating. Small holes should be punctured in the lid for air circulation and the containers stored in a well ventilated room. Microworm can be harvested daily for about 28~56 days using the same culture medium. However, it largely depends on the culture medium used. It is a good to have at least two cultures running at the same time. The second culture can be started about two weeks after the first. A culture will sometimes rapidly decline in production of worms. Having a second culture in production will ensure availability of worms at all times. Culture medium can be prepared from almost any grain flour, yeast, and water. However, research has shown that the type of culture medium used has a great influence on worm yields. Trials conducted using three mediums - wheat flour, oatmeal, and cornmeal proved that yield of worms in wheat flour was significantly greater than in oatmeal or cornmeal. Production of worms stopped after day 20 in cornmeal, day 33 in oatmeal, and day 53 in wheat flour. The addition of yeast during initial media preparation was found to have no effect on worm yields. However, the addition of yeast on a weekly basis to the wheat flour medium gave a significant greater yield of worms than did untreated wheat flour. The wheat flour was mixed with water to form a smooth paste and placed in the culture container. After inoculation with live worms, the addition of 5 ml of a yeast solution, consisting of 7 gm bakers or brewers yeast (*Saccharomyces cerevisiae*) dissolved in 70 ml water; was lightly sprayed over the medium every 7 days. The addition of yeast will also inhibit the growth of nematophage fungi.

Another method is using oatmeal (porridge), using one part oats with one part of water. Place the mixture into the culture container and spread to a thickness of 15~25 mm and microwave for three minutes. The mixture is then allowed to cool to room temperature. Any media on the sides of the container should be removed with a damp cloth. After the mixture has cooled, place the starter culture on top of the porridge. Within 3-6 days we can observe the surface moving and on using a magnifying glass, hundreds of tiny worms can be observed. The production of worms can be increased by sprinkling dry yeast powder over the surface of the mixture. Adding the yeast once a week should be sufficient. If the culture medium becomes very watery, add a slice of bread to the container to soak up the moisture. The addition of bread has a similar effect as the bakers yeast.

In the third method only a slice of white bread and brewers yeast is required. This culture method produces the best results in terms of the number of worms produced. First, cut the crusts off the slice of bread and place it on the bottom of the container. Mix 5 g of brewers' yeast with ¼ cup of water and pour the mixture onto the bread, making sure that the bread is completely saturated. It is important that there should be very little excess fluid in the container when the container is tilted. Next add the starter-culture of microworm, by spreading it over the surface of the bread. Replace the container lid securely, and place the container in a warm area.

Within three to four days, the culture will be thriving with worms migrating up the side of thecontainer. As the bread is consumed another slice can be added to keep the culture active. Afterthe addition of around 3 or 4 slices of bread the culture needs to be replaced. If the culture becomes too wet, more bread should be added to absorb the excess moisture.

NB: Remember the wetter the culture, the lower the production of worms. Care should be taken not to allow the entry of mosquitoes or house fly into the media to avoid contamination

NB : Do not dip your finger or brush into the culture medium to collect worms, as any culture media residues should be minimised in order to avoid pollution of the aquarium water.

Another method of harvesting is to lay wooden ice sticks (or similar objects) on the surface of the culture. The worms will crawl onto the sticks and the worms can then simply wash the stick in the aquarium water. Another method is to using a damp (thick) paper towel/tissue paper cut to fit over about half of the culture surface. To harvest the worms a spoon or spatula can be used to gently scrape the worms off the paper towel, without tearing the towel (Wedekind, 2008).

Important Points to Observe

- ☆ Uneaten worms will die and pollute the aquarium water, particularly in a small aquarium.
- ☆ If left unattended, it can eliminate an entire batch of fish larvae in a matter of hours.
- ☆ To prevent this problem, feed the larvae three or four times per day in small amounts rather than one or two large quantities.
- ☆ Microworms offers a cheap, simple and nutritious feed for feeding the larvae of most freshwater fish species. It is also suitable for feeding juveniles and adults of fishes.

- ✰ Microworm can be fed alone or in combination with other feeds such as brine shrimp nauplii, rotifers, zooplankton, egg yolk and dry diet.
- ✰ A feeding program utilizing a combination of feed items is better able to meet the nutritional requirements of all freshwater fish larvae.

Enrichment of Microworms

Studies on enriched media for microworm have shown encouraging results. The nutritional quality of microworm can be enhanced by the use of the direct enrichment technique. Enrichment is simply carried out by adding the product to the culture medium. The microworm were cultured in two media one with oatmeal and the other with spirulina enriched oatmeal, in 15x15x5 cm plastic containers with 200g oatmeal and 300 millilitres (ml) purified water. Five grams of spirulina was used in the medium. The results show that growth of the microworm population in the spirulina-enriched medium presented the highest abundance of individuals on the second week of culture, whereas the population grown in the oatmeal medium showed the highest abundance on the fifth week of culture but did not reach the number of organisms attained by the population cultured in the spirulina-enriched medium. (de Lara *et al.*, 2007).

The amino acids content of the populations from both media were compared to those reported for brineshrimp fed with spirulina, observing that the amounts were higher for most amino acids in microworm cultured in the spirulina-enriched medium. The composition of fatty acids in the microworm cultures in both media depicted significant differences for the linoleic, arachidonic, and eicosapentaenoic fatty acids, which were found in a higher percentage than reported for microworm cultures in oatmeal supplemented with sunflower oil. This information shows that spirulina accelerates growth of microworm populations and allows the presence of amino and fatty acids. On using a culture medium which was fortified with a 10 per cent fish oil emulsion, obtaining nematodes that had significantly higher total lipid content and elevated levels of highly unsaturated fatty acids (HUFA) (Rouse *et al.*, 1992). Additional investigations concerning the effect of an added oil source (fish oil or sunflower oil) on body composition, average yields and multiplication factors on the nematodes were conducted by Schlechtriem *et al.* (2004). Corn oil and yeast were added in varying combinations to culture media of *Panagrellus redivivus* to effect a change in growth and fatty acid content. The addition of corn oil or yeast to cultures increased nematode growth over standard media. Combinations of corn oil plus yeast increased nematode growth by 68 per cent.

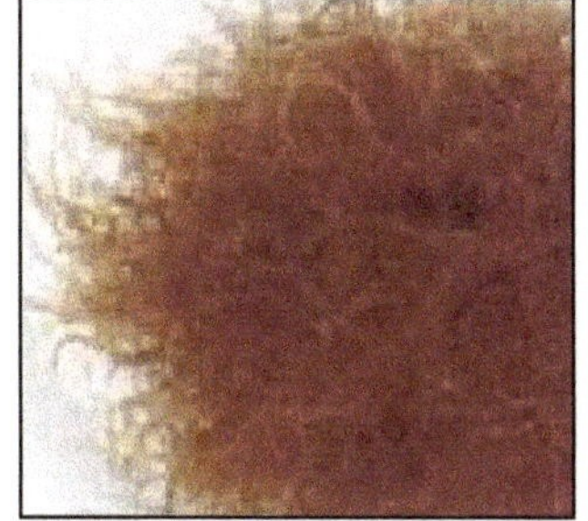

Figure 10.6: Tubifex Worms.

Tubifex Worms

Tubifex worms are another type of live feed normally used to feed different types of freshwater aquarium fish. It is often used to induce breeding especially in goldfish and the tubifex worms culture can be easily bought at any aquarium fish stores. Before feeding the tubifex worms to the fish, one must take extreme caution in order to ensure that the worms are properly cleaned and washed. Leave the worms in a container of clean fresh water and then change the water every 4-5 hours in order to ensure that contaminants are flushed out. Without feed, these worms can last up to a day or two so it might be a good idea to change the water frequently within this period before feeding to your fish.

Bloodworms

Figure 10.7: Bloodworms.

Bloodworms are really the larval form of one of the chironomid midges - insects rather than worms. Fishes love the taste of blood worms. The hemoglobin that colors them red is an excellent iron source. Full-grown blood worms grow to about an inch. Their size makes them ideal for two to six-inch fish. Adults resemble very much like mosquitoes, the adult flies lay their eggs in water and the hatched out larvae are the bloodworms. Adults live three to five days. On observing carefully under the water we can see for blood worm eggs on sticks and grasses near the muddy water. They look like tiny dots, very much like pond snail eggs. These eggs can be collected and cultured. They hatch in 24 to 48 hours.

Negative phototropic blood worm larvae avoid the light. They can probably attain faster growth by putting black plastic sheets over them. They construct tiny tubes in which they live during the day. This makes them difficult to harvest. It is easy to harvest them at night, when they emerge from their tubes. The worms can also be easily collected by cutting off their aeration which also forces them to come out.

Feeds

Blood worms eat "micro feeds." It can be raised on chicken manure, horse manure, vegetable wastes and other waste products. Partial water changes may be necessary during culture period.

Feeding Schedule

Sprinkle powdered oats on top. Monitor and feed every two to six days. to help keep track of their size. Feed small amounts. Aeration can be provided. Water good, old aquarium water or de-chlorinated tap water can be used. Avoid extremes fluctuations in pH levels.

Harvesting

Harvest the worms at night. Turn off aeration and net the worms. Excess worms can be frozen in the fridge. Blood worms are one of the favorite live feeds for discus fish. Some are sold as freeze-dried worms, stored in refrigerator and sold in packets. However, the blood that comes with it dissolves easily in aquarium water, therefore after the feeding schedule; it has to be followed up with water change. Bloodworms are preferred among fish owners for its high nutritional value and protein content. Discus fish fed with bloodworms often exhibit bright colouration and attain fast growth.

Fairy Shrimp (*Streptocephalus* sp.)

The fairy shrimp are freshwater relatives of the brine shrimps, *Artemia* spp. Fairy shrimp nauplii closely resemble brine shrimp nauplii and are similar in size. Fairy shrimp normally inhabit temporary ponds that do not have fish because they are fed upon by fish in natural waters. Fairy shrimp have the potential to be used as a feed item for ornamental fishes that benefit from live feed.

The advantages of using the freshwater anostracans as live feed includes:

- ✰ High individual biomass
- ✰ Possibility of bioencapsulation with PUFA
- ✰ Rich source of carotenoid pigments
- ✰ Enrichment and bioencapsulation. The nutritional quality of the fairy shrimps can be improved by enriching them with exogenous source of nutrients.

Bioencapsulation is defined as the process by which live feed organisms are enriched with specific nutrients or drug molecules (for example, vaccines) and fed to the target organism.

Enrichment of live feed with n3 highly unsaturated fatty acids (n3-HUFA), particularly eicosapentenoic acid (EPA) and docasahexenoic acid (DHA) has proven to be beneficial in larval and maturation feeds. Rotifers and Artemia are naturally deficient in n-3 HUFA and particularly DHA. This can be corrected by pre-feeding them with n3 HUFA enriched products.

White Worms

White worms are very good fry - feed that can be cultured at backyard. Some moist and nutritious garden soil is required as substrata for the white worms to grow. Fill 75 per cent of a shallow box with garden soil. Water the soil until it is quite damp. It is important that the soil should never be allowed to dry out and also water logging of the soil should be prevented. Add the white worm culture to the soil together with a few small pieces of moist white bread. Cover the box with a lid.

Figure 10.8: White Worms.

The box or the lid must have some air holes to allow for ventilation. Place the box in a dark place where the temperature is around 16-18°C. New bread should be added to the box every 3-4 days and make sure that the soil is constantly kept moist. Any uneaten feed should be removed before it decomposes.To collect white worms for the fish, some soil from underneath the bread has to be removed. It will be seen that naturally the largest congregations of white worm will be right beneath the bread pieces. Pour the soil with the white worms into a bowl or dish filled with water. When the worms have become separated from the soil they can easily be collected and dropped directly into the aquarium.

Preparation of Frozen Live Feeds

- ✰ Collect the live feed using, an aquarium net or a spoon to scoop the worms with debris.
- ✰ The debris should be thoroughly rinsed keeping the above in a bowl type netting under fast flowing tap water. There should not be any dirt in the cubes.
- ✰ Drain the live feed.
- ✰ After the live feed is rinsed and drained transfer those into the empty ice cube trays.
- ✰ Add water to the ice cube trays to desired height. Use bottled or boiled water that has been cooled. This water must be clean because it will eventually be added later to the fish tank.
- ✰ Place the ice trays into the freezer overnight or until the ice cubes are frozen solid.
- ✰ Once frozen, carefully remove the ice cubes from the ice trays.
- ✰ Place frozen fish feed - ice cubes into plastic resealable bags. Seal the bags tightly and write the date when the frozen fish feed cubes were made on the plastic bag.
- ✰ Always keep the frozen fish feed cubes in the freezer
- ✰ Use this frozen feed before one year.Frozen Feed that is older than a year has to be discarded.
- ✰ 1 frozen cube can be fed directly into the fish tank every alternate day. Alternate frozen feed with other types of feeds.

NB : Many types of live feed can be added into the cubes at the same time. Bloodworms, brine shrimp and *Daphnia* can be frozen together in one cube.

Rotifer Cultures

Diets Used in Rotifer Cultures

Algae

Marine micro-algae are the best diet for rotifers and very high yields can be obtained if sufficient algae are available. The most common algae used in rotifer cultures is *Nannochloropsis oculata* (Lubzens, 1987; Hirayama *et al.*, 1989; Fukusho, 1989) with a size of 2-3 mm in diameter and a relatively high content in 205n-3 fatty acid (EPA), *Tetraselmis tetrahele* or *T. suicica* which have a cell diameter of 20-30 Im and high EPA content, *Isochrysis galbana* containing high level of 226n-3 fatty acid (DHA). Some other micro-algae including *Dunaliella tertiolecta, Pavlova lutheri,*

Chlorella sp. and *Stichococcus* sp. have also been used as feed for rotifer cultures. Microalgae are believed to play a role in stabilizing the water quality, nutrition of the larvae and microbial control. Apart from their high nutritional value, some other advantages can be obtained from the microalgae as feed for rotifers. Algae act as bacteriostatic, controlling bacterial development. Algae act as water conditioner, controlling the water quality of the medium and oxygenating the water through the photosynthesis process.

A stable micro algae supply for mass production of rotifers is difficult to obtain. Therefore, alternatively, baker's yeast is commonly being used. Baker's yeast was used as feed for rotifers. They reported that the rotifers could grow on a mixed feed (50 per cent *Chlorella* and 50 per cent baker's yeast) as well as with 100 per cent *Chlorella*. There are several yeasts that can be used as rotifer feed, that is baker's yeast (fresh and instant) (*Saccharomyces cerevisiae*), caked yeast (Rhodotorula) and marine yeast (*Zygosaccharomyces marina, Torulopsis candida var. marina, T. larvae* and *Saccharomyces acidosaccharophill*). Baker's yeast has been used as a suitable algal substitute for *Brachionus* (Hirayama, 1987), because of its small particle size of 5-7 Im in diameter, high content of protein and also the presence of bacteria growing on the yeast surface. Although yeasts have been accepted as feed for rotifer cultures, they contain very low concentration of long chain highly unsaturated fatty acids (HUFA) of the n-3 series, mainly 205n-3 and vitamin B12.

Culture Techniques

(*a*) Batch Culture

Batch culture systems seem to be the most common type of rotifer production used in hatcheries. The size of the rearing tanks varies from 500 to 1000 l plastic tanks up to 10000 l for concrete tanks. In these systems the rotifers are inoculated at a density of 50 to 200 rotifers.ml^{-1}. The density at harvest time is about 600 rotifers. ml^{-1} after 4 days culture (Sorgelooos and Lavens, 1996). In the batch culture technique rotifers are harvested completely. A part of the harvested rotifers are administered as feed for fish larvae or crustaceans and the remaining is used as inoculum for the next culture with a density of 250 rotifers.ml^{-1}. Depending on the culture volume and rotifer density during the rearing period, two strategies can be applied a constant culture volume can be maintained with increasing rotifer density or the volume of the culture can be adjusted in order to maintain a constant rotifer density.

(*b*) Semi-Continuous Culture

In high density cultures of rotifers, a large amount of suspended organic matter accumulates in the culture medium. Such wastes are mainly composed of rotifer feces, amictic egg shells, microbes (bacteria, protozoa, fungi, etc.), the feed organism *Chlorella* and flocks of various sizes are formed and coagulate (Yoshimura, 1997a). In order to avoid this phenomenon of self-pollution, the semicontinuous culture system has been developed. In the semi-continuous culture systems the rotifer density is kept constant by harvesting periodically. Girin and Devauchele (1974) removed about 25 per cent of the culture volume every day and replaced it by the same amount of new water. The system is therefore also known as the thinning culture method. If all requirements of this system are satisfied the method allows the maintenance of a stock of a constant number of individuals (Coves *et al.*, 1990). Semi-continuous culture systems are usually performed in larger tanks (50-200 m^3) than the ones used in the batch culture. The culture period is longer than that in the batch culture system. Morizane (1991) reported that they could continue culturing rotifers without changing tanks, harvesting a large number rotifers, for an entire year. The inoculated density of rotifers is from 50 to 200 rotifers.ml^{-1} and can reach up to 300 to over 1000 rotifers.ml-1 in 3 to 7 days at harvesting time, using micro-algae and baker's yeast.

(*c*) Continuous Culture

Continuous culture systems are a logic process in the intensification of rotifer production. The aim of this system is nearly the same as for the semi-continuous culture system, to maintain good water quality by improvement of water management through high water exchange rate and the use of chemostats. The new water is always supplied into culture tanks, so that the water quality is kept in good quality or nontoxic without the need for any procedure such as pH control for reducing unionized ammonia. In this system, a constant rotifer density with high quality is reached, and it is also possible to maintain the culture without any decline of rotifer productivity for a long period. Abu-Rezeq (1997) reported that the continuous culture systems have higher productivity than batch and semi-continuous culture systems. The initial density of rotifers varies, and during the culture period the rotifer density is maintained constant and the production is dependent on other factors such as feeding regime and water quality. Although the continuous culture systems have a lot of advantages they are only applied on an experimental scale and are not applied in the hatcheries. Since this system is very costly and a lot of variables need to be controlled the risk for technical failure is considerably increased.

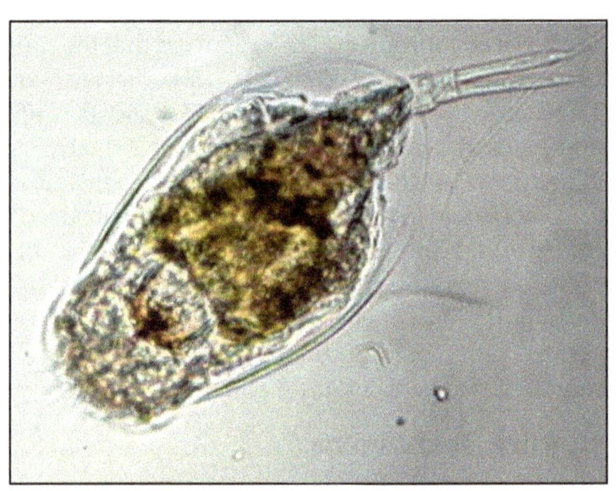

Figure 10.9: Rotifer (*Brachionus* sp.).

(*d*) Ultra-high Density Culture

The intensive ultra-high density rotifer culture techniques have been firstly developed by Japanese scientists. Yoshimura *et al.* (1995) reported that very high rotifer productions could be achieved in a 1 m^3 tank in a batch culture method in 2-day intervals with a initial density of 10,000 individuals.ml^{-1}. The latest, ultra-high density (maximum density from 20,000 up to 40,000 rotifers.ml^{-1}) rotifer mass culture has been developed based on concentrated freshwater *Chlorella* as feed (Yoshimura *et al.*, 1994, 1997a, Fu *et al.*, 1997). The ultra-high density culture systems are an effective way to produce rotifers without expanding the culture space. These systems have several advantages.

- ☆ Much lower labour and space needed
- ☆ High production of rotifers
- ☆ Consistent or year-round production

In the rotifer mass production system the most labour-intensive step is the harvesting of the culture tanks before feeding or enrichment (Dehasque *et al.*, 1997). Several advantages can be obtained from the automated system over the manual system are as follows:

- ☆ Production techniques are simplified
- ☆ More intensification is made possible which means less tanks are needed and required space/infrastructure is reduced
- ☆ Less labour is required
- ☆ Manipulations are reduced and higher outputs per units of volume are reached
- ☆ The extra cost to install automated procedures are minimal (concentrator/rinser; pumps) and compensated by reduction in tanks and labour

Enrichment with Nutrients

a) With Oil Emulsions

For the enrichment or boosting of rotifers several approaches can be followed i) the adjustment of the lipid and vitamin content of the rotifers just before feeding them to other organisms is referred to as short term enrichment (generally less than 8 h exposure) and ii) the feeding of rotifers on a complete diet or long term enrichment (rearing of the rotifers on the enrichment diet for more than 24 h).

b) With Vitamins

The vitamin C content of rotifers reflects the dietary ascorbic acid (AA) levels both after culture and enrichment. Rotifers cultured on *e.g.* instant baker's yeast contain low AA levels (150 mg.g^{-1} DW), while the AA content in Chlorella-fed rotifers may vary (from 1000 up to 2300 mg.g^{-1} DW) depending on the quality of the algae. Oil-soluble vitamins or derivates from water soluble vitamins (ascorbyl palmitate (AP) for ascorbic acid) have been formulated in the commercial lipid enrichment products. The non-bioactive ascorbyl palmitate is accumulated by the rotifers together with the oil emulsion and converted to free AA by the enzymes of the rotifers. Merchie *et al.* (1995) demonstrated that this process was very effective since 5 per cent AP (w/w) in the emulsion produced rotifers with an active AA concentration of 1700 mg.g-1 DW after 24 h enrichment and this high concentration remained in the rotifers after storage in seawater during the next 24 h.

c) With Proteins

The protein content of rotifers is reported to vary between 28-67 per cent of dry weight, whereas the amino acid profiles of rotifers fed different diets appear to be fairly constant and independent of feed quality. A positive effect was seen on growth and survival when fast reproducing rotifers (*i.e.* high protein content with high protein/lipid ratio) were fed.

References

Cheng-Sheng Lee, Patricia J. O'Bryen. and Nancy H. Marcus. 2005. Copepods in Aquaculture. Blackwell Publishing.

Carl D. Webster and Chhorn Lim. 2002. Nutrient Requirements and Feeding of Finfish for Aquaculture. CABI Publishing.

de Silva, S.S. and Anderson, T.A. 1995. Fish Nutrition in Aquaculture. Springer.

Felicity Huntingford, Malcolm Jobling and Sunil Kadri. 2012. Aquaculture and Behavior. Wiley Blackwell Ltd.

Hagiwara, A., Snell, T.W., Lubzens, E. and Tamaru, C.S. 1997. Live Food in Aquaculture. Kluwer Academic Publishers.

Hertrampf, J.W. and ý Piedad-Pascual, F. 2000. Handbook on Ingredients for Aquaculture Feeds. Kluwer Academic Publishers.

Jean Guillaum. 1999. Nutrition and Feeding of Fish and Crustaceans. INRA, IFREMER.

Joan Holt, G. 2011. Larval Fish Nutrition. Wiley Blackwell Ltd.

Josianne and ýLesley McEvoy. 2003. Live Feeds in Marine Aquaculture. Blackwell Science Ltd.

Chapter 11

Ornamental Fish Nutrition: Scenario and Strategies

Cheryl Antony

Institute of Fisheries Technology,
Tamil Nadu Fisheries University,
Chennai – 51

The nutritional requirement of fish change with species, size, age and season. Fish are unique in that only a small proportion of species use plant material as their primary feed source. Those species, such as tilapia, grass carp and stonerollers, consume nearly their body weight in vegetation daily. Digestion of plant materials by these species is poor. Most fish eat a diet rich in protein and fat. Protein is used for both tissue growth and energy. Therefore, most commercial fish diets contain > 30 per cent protein, and diets for very young fish may contain nearly 60 per cent protein. Commercial diets are partially cooked during production, which increases the digestibility of plant products. The digestibility of plant and animal proteins appears similar in commercial diets, but carp were able to extract more energy from the animal diets (Takeuchi *et al.*, 2002).

Lipids are also an important energy source for fish, and a source of essential fatty acids. The most desirable fats are unsaturated. Unsaturated fats contain less hydrogen, so they are more easily metabolized by the fish. Diets containing high levels of unsaturated fats have a short storage life, because they oxidize and become rancid when exposed to the atmosphere. Rancid diets are unpalatable to fish, so diet intake decreases, which causes an increase in direct loading to the filter system unless feeding rates are decreased. Goldfish do well with diets that contain 3-6 per cent lipids. Not surprisingly, chironomid (midge) larvae, the preferred feed item of ornamental fishes, contain 48 per cent protein, 14 per cent lipid, and 23 per cent carbohydrate, with 460-610 kcal/kg dry weight. The type and quantity of feed required to keep the fish in good health varies with age, size and water temperature. The natural diet of newly hatched goldfish and koi is dominated by zooplankton, such as rotifers and daphnia. Zooplankton is a rich source of protein and calories. Protein content of zooplankton varies from 30 – 50 per cent depending on life stage and nutrient availability of their phytoplankton diet. The protein content in zooplankton varied from 50 – 71 per cent. Accordingly, larval goldfish were found to grow best on prepared diets containing about 50 per cent protein. Small fry goldfish and koi grew best on prepared diets containing about 40 per cent protein. As the fishes become mature their nutritional requirements diverge. Adult goldfish require only 29 per cent protein and they will continue to grow when receiving only 1 per cent body weight of a diet containing 36 per cent protein. Koi carp, on the other hand require 30-35 per cent protein, and N excretion was reduced when dietary protein content was 35 per cent indicating more efficient nutrient use with the higher protein diet.

Commercial diets for goldfish and koi are available in a range of protein percentages. In general, diets with at least 30 per cent protein will yield acceptable growth of ornamental fishes. Diet digestibility will vary according to the primary nutrient sources. Goldfish can accommodate up to 70 per cent carbohydrate in the diet, while koi have little ability to digest carbohydrates. Since koi do not digest carbohydrate well it is important to use the most digestible carbohydrate-based protein sources. Koi digest wheat germ better than corn, and corn better than rice bran.

Fish have different dietary nutrient requirements based on size, but there are no such restrictions based on water temperature. However, fish metabolism is based on water temperature so the amount of feed given to fish should vary accordingly. During peak growth periods, goldfish and koi can consume about 1.5-2.0 percent of their body weight as dry feed each day. As the water cools in fall, fish appetite also decreases. Since goldfish and koi are ornamental fish and valued for their coloration, size and fin development, manufacturers of prepared diets developed color-enhancing formulas to promote brilliant colors in the fish. Color enhancing diets do work, but their effectiveness varies with variety of fish, fish age and genetics. Development of red coloration in oranda goldfish was enhanced by feeding diets containing astaxanthin, lutein and zeaxanthin for 20 weeks. Water quality and temperature can also affect the color of fish. Coloration is often intensified in cool, clear water, and during spawning season.

Carotenoids in Ornamental Aquaculture

A variety of carotenoids, both synthetic and naturally occurring products or are being developed for use in aquaculture. Included are synthetically produced astaxanthin (3,3'-dihydroxy-P, P-carotene-4,4'-dione) and canthaxanthin (beta-carotene-4,4'-dione) and natural materials such as krill, Spirulina, crustacean-meals, marigold, Capsicum, and other xanthophyll-containing vegetable meals. Added to this are commercially available products of the astaxanthin-rich yeast *P h f i* rhodozyma. Another microbial source being considered is the microalga *Haemafococcus pluvialis* Carotenoids are the primary source of pigmentation in ornamental or tropical fish, responsible for various species-related yellow, red, and related colors. Normally these are obtained through carotenoid-containing organisms in the aquatic feed chain, but commercial feed ingredients such as yellow corn, corn gluten meal, and alfalfa are used as sources of carotenoids such as zeaxanthin and lutein. Other carotenoid-rich ingredients used are marigold meal (lutein), red pepper (*Capsicum* sp.) extract (capsanthin) and krill or crustacean meals (astaxanthin). Canthaxanthin has been shown to be an effective pigment for the tropical fish *Trichogaster leeri*, especially to enhance maturation.

Biological Activities of Carotenoids in Aquatic Species

Increasing supportive evidence has been observed regarding the biological role of astaxanthin in fish metabolism other than that of a pigmenting agent. Carotenoids in general enhance both the nonspecific and specific immune system. Proposed functions include protection against UV light, serving as provitamin A, enhancing tolerance to elevated ammonia levels and low oxygen levels, stimulation of growth, maturation rate, fecundity rate as a fertilization hormone, and improvement of egg quality.

Astaxanthin also is a strong inhibitor of lipid per oxidation. The activity of astaxanthin in protection of biological membranes from oxidative injury by inhibition of mitochondrial lipid peroxidation has been studied. It was shown that astaxanthin functions as a potent antioxidant both in vivo and in vitro. Astaxanthin protects biological membranes from oxidative injury by inhibiting mitochondrial lipid peroxidation. Astaxanthin absorption also depends on the concentration used and whether it is provided in its free form or as a diester.

Fat-Soluble Carotene and Fish

Fat-soluble carotenoid is responsible for the bright hues in some fish. Krill and brine fish are some of the feeds that are rich in pigments.

Protein

Protein is the single most important nutrient that the fish needs to grow. On a dry-weight basis, this makes up the maximum weight in their body structure. Amino acids are derived from proteins and the fish uses them to make new body tissues as well as enzymes. Fish are very adept at converting feed to body tissues. That is why fish need lesser amounts of feed than do most other animals. Carbohydrates are almost non-existent in the feed intake for many fish species, since energy is also derived from proteins.

The quantity of protein required for the fish to be healthy depends on a number of variables like the species of fish, amount of natural feed available, growth rate, etc. Fry and larvae require a more protein rich diet to maximize their adaptability and chances of survival. As the fish grow larger, their dependency on protein reduces. The temperature of the water also affects protein requirements. Fish meal, soybean meal, fish hydrosylate, skim milk powder, legumes, and wheat gluten are excellent sources of protein. Additionally, the building blocks of proteins (free amino acids) such as lysine and methionine are commercially available to supplement the diet.

Fatty Acids

Fatty acids are sources of energy for most fish. Fish that live within the confines of an aquarium are naturally prone to obesity. They do not use up their excess energy in swimming long distances or looking for feed. In most cases, excess fat can be damaging to the general health of the fish. Some fish lose their reproductive capabilities if there is too much body fat.

Carbohydrates

The carbohydrates, which include starches, sugars, cellulose and gums containing only the elements carbon, hydrogen and oxygen which are usually the cheapest source of energy in foods and feeds. Fish and shrimp, however, vary in their ability to digest carbohydrate effectively. Many fish appear to be able to utilize simple carbohydrates, such as sugars, more effectively than complex starches; the reverse appears to be true for shrimp and prawns.

This observation may be confused by the beneficial effect that carbohydrates tend to have on the structural integrity of the feed, caused by the binding quality of starches. Carnivorous fish such as salmon and trout and particularly marine fish are not efficient converters of carbohydrate. Channel catfish, like shrimp, appear to be able to utilize complex carbohydrates more readily than simple sugars. Channel catfish and carp can utilize quite high levels of dietary carbohydrate. The natural diet of grass carp is very high in this component.

Some carbohydrates are normally regarded as indigestible. These are reported in feed composition as 'fibre' or 'crude fibre'. Fibre includes substances such as celluloses (from plants), lignin, chitin, etc. Many fish do not have the enzyme cellulase which is necessary for the digestion of cellulose, and fibre is usually regarded as unavailable as an energy source. At small levels, however, it may aid pelletability. Cellulase however is produced by the gut bacteria of many fish, as is chitinase in crustacea, and herbivorous fish are able to digest fibre.

Vitamins and Minerals

Vitamins are vital to fish health. These are organic substances that act as catalysts for many of the biochemical reactions within the fish. Almost all vitamin deficiency will increase the fish's susceptibility to diseases and stress. The best way to get a rich supply of vitamins to the fish is to feed small quantities of diverse feed. Providing frozen or fresh vegetables and live feed can also supply the much-required vitamins to your fish.

Minerals are also necessary for fishes. Bones, teeth and scale tissues require lots of minerals. The minerals also carry out many supportive functions. The variety and amount of vitamins and minerals are so complex that they are usually prepared synthetically and are available commercially as a balanced and pre-measured mixture known as a vitamin or mineral premix. This premix is added to the diet in generous amounts to ensure that adequate levels of vitamins and minerals are supplied to meet dietary requirements. Though mineral supplements will help to compensate this deficiency from feeds or the environment, excess of some minerals can be poisonous. Therefore, mineral supplements should not be used indiscriminately.

Pigments

A variety of natural and synthetic pigments or carotenoids are available to enhance coloration in the skin of freshwater and marine ornamental fish. The pigments most frequently used supply the colors red and yellow. The synthetically produced pigment, astaxanthin (obtained from companies such as Cyanotech and F. Hoffmann-La Roche Ltd.), is the most commonly used additive (100-400 mg/kg). Cyanobacteria (blue-green algae such as *Spirulina*), dried shrimp meal, shrimp and palm oils, and extracts from marigold, red peppers and Phaffia yeast are excellent natural sources of pigments.

Binding Agent

Another important ingredient in fish diets is a binding agent to provide stability to the pellet and reduce leaching of nutrients into the water. Beef heart has traditionally been used both as a source of protein and as an effective binder in farm-made feeds. Carbohydrates (starch, cellulose, pectin) and various other polysaccharides, such as extracts or derivatives from animals (gelatin), plants (gum arabic, locust bean), and seaweeds (agar, carageenin, and other alginates) are also popular binding agents.

Preservatives

Preservatives, such as antimicrobials and antioxidants, are often added to extend the shelf-life of fish diets and reduce the rancidity of the fats. Vitamin E is an effective, but expensive, antioxidant that can be used in laboratory prepared formulations. Commonly available commercial antioxidants are butylated hydroxyanisole (BHA), or butylated hydroxytoluene (BHT), and ethoxyquin. BHA and BHT are added at 0.005 per cent of dry weight of the diet or no more than 0.02 per cent of the fat content in the diet, while ethoxyquin is added at 150 mg/kg of the diet. Sodium and potassium salts of propionic, benzoic or sorbic acids, are commonly available antimicrobials added at less than 0.1 per cent in the manufacture of fish feeds.

Attractants

Other common additives incorporated into fish feeds are chemoattractants and flavorings, such as fish hydrolysates and condensed fish solubles (typically added at 5 per cent of the diet). The amino acids glycine and alanine, and the chemical betaine are also known to stimulate strong feeding behavior in fish. Basically, attractants enhance feed palatability and its intake.

Other Feedstuffs

Fiber and ash (minerals) are a group of mixed materials found in most feedstuffs. In experimental diets, fiber is used as a filler, and ash as a source of calcium and phosphorus. In practical diets, both the ingredients should not be higher than 8-12 per cent of the formulation. A high fiber and ash content reduces the digestibility of other ingredients in the diet resulting in poor growth of the fish.

The addition of fish or squid meal will enhance the nutritional value of the diet and increase its acceptance by the fish. Fresh leafy or cooked green vegetables are often used. Although vegetables are composed mainly of water, they contain some ash, carbohydrates and certain vitamins. Greens are examples of relatively nutritious vegetables. Feeds are formulated to be dry, with a final moisture content of 6-10 per cent, semi-moist with 35-40 per cent water or wet with 50-70 per cent water content. Most feeds used in intensive production systems or in home aquaria are commercially produced as dry feeds. Dry feeds may consist of simple loose mixtures of dry ingredients, such as "mash or meals," to more complex compressed pellets or granules. Pellets are often broken into smaller sizes known as crumbles. The pellets or granules can be made by cooking with steam or by extrusion. Depending on the feeding requirements of the fish, pellets can be made to sink or float.

Ingredients of Quality Fish Feed

Fish feed should ideally provide the fish with fat (for energy) and amino acids (building blocks of proteins) and the fish feed (whether flake or pellet) must be digestible in order to prevent build up of intestinal gas, renal failure and infections (such as swim bladder problems and dropsy) and to avoid aquarium pollution due to excessive ammonia. Aquatic diets for carnivores must contain vegetable matter such as spirulina.

Building Block Ingredients of Fish Feed

- Amino acids are the basic components of proteins. An example of an aquatic diet that is a good source of amino acid is a crumbled hard boiled egg offered to small fry. Large amounts of DL-Methionine enhance the growth of the head of the Lionhead goldfish.
- Fats that are broken down into fatty acids are the main source of energy in fish especially for the heart and skeletal muscles. Fats also assist in vitamin absorption. Vitamins A, D, E and K are fat-soluble or can only be digested, absorbed, and transported in conjunction with fats.
- Carbohydrates are molecular substances that include sugars, starches, gums and celluloses. Most of the carbohydrates that are incorporated into aquatic diets are of plant origin and are sources of the enzyme amylase. Carbohydrates, however, are not a superior energy source for fish over protein or fat but digestible carbohydrates do spare protein for tissue building. Unlike in mammals, glycogen is not a significant storage depot of energy in fish.

Basal Feed Ingredients

- Fish meal (protein source) have two basic types: (a) those produced from fishery wastes associated with the processing of fish for human consumption and (b) those from specific fish (low value fishes) which are harvested solely for the purpose of producing fish meal.
- Shrimp meal is made from shrimp wastes that are being processed before freezing or from whole shrimp that is not of suitable quality for human consumption. The material to be made into shrimp meal is dried (sun-dried or by using a dryer) and then ground. Shrimp meal is a source of pigments

that enhances the desirable color in the tissues and skin of fish. It is also a secondary supplemental protein source for fish.

- ☆ Squid meal is made from squid viscera portions from cannery plants including the eggs and testis. Squid meal is a highly digestible protein source for fish which provides a full range of amino acids, vitamins, minerals and cholesterol (1.0–1.5 per cent) of cholesterol suitable for fish fry and young fish.
- ☆ Brine shrimp (adult *Artemia*) is a common feed source for fish that are available in adult-form, as eggs or freeze-dried. Brine shrimp is a source of protein, carotene (a color enhancer) and acts as a natural laxative in fish digestive systems. Brine shrimps can also supply the fish with vegetable matter due to their consumption of algae.
- ☆ Soybean meal is a high protein source for fish and has become a substitute for traditionally-used marine animal meals.
- ☆ Spirulina is a blue-green plant plankton rich in raw protein, vitamins A1, B1, B2, B6, B12, C and E, beta-carotene, color enhancing pigments, a whole range of minerals, essential fatty acids and eight amino acids required for complete nutrition.
- ☆ Whole wheat (carbohydrates) is not the best source of energy in fish but is an excellent source of roughage for fish such as Goldfish and Koi. It is also a natural source of vitamin E which promotes growth and enhances coloration.

Different Types of Artificial Feeds

Most commonly, fish feed can be divided into 3 main categories:

1. *Manufactured feed*: This includes floating and sinking pellet, granular and flake feed.
2. *Freeze-dried feeds*: Worms, larvae, brine shrimp and krill.
3. *Live feeds*: Maggots, fresh insect larvae, live worms, and brine shrimp.

Dry Feeds

Flake feed is a type of proprietary or artificially manufactured fish feed consumed by a wide variety of tropical and saltwater fish and invertebrates. It is ideally suited to top dwellers and mid-water fish though numerous bottom dwelling species consume flake feed once it has settled on the bottom. If the aquarium consists of bottom dwellers mainly, it would be good to pre-soak the flake feed so that it will sink to the bottom as soon as it is introduced into the water.

Flake feed is baked to remove moisture and create the flaking, thus allowing for a longer shelf life. Flake feeds providing for all the fish's nutrient needs. They are an excellent source of various minerals too. They are easy to keep and have fairly long shelf lives. They come in various flavors. Flakes are another form of dry feed and a popular diet for aquarium fishes. Flakes consist of a complex mixture of ingredients, including pigments. These are made into a slurry which is cooked and rolled over drums heated by steam.

NB: *Drop a pinch or two of the flake into the water and observe the rate at which the feed is eaten. Then decide how much quantity needs to be fed in the next feeding.*

Generally the more moisture a particular example of fish feed contains, the more readily it will deteriorate in quality. Dry feeds are available as pellets, sticks, tablets, granules, and wafers, manufactured to float or sink, depending on the species to which it is feed.

Vacation Feed

Vacation feeds — also known as "feed blocks" — are designed to be placed inside the aquarium to forgo feeding on the absence of the owner. These blocks release small amounts of feed as they dissolve. Feed blocks can be a good choice for smaller tropical fish, but can pollute the water.

Medicated Fish Feed

Medicated fish feed is a safe and effective method to deliver medication to fish. One advantage is that medicated feed does not contaminate the aquatic environment and also, unlike bath treatments, does not negatively affect fish, filtration and algal growth in the aquarium. The parasites will get treated on spot by medicated feed, because the fish is ingesting the feed.

Freeze-Dried Fish Diets

Freeze-dried and frozen fish feeds were primarily developed for tropical and marine fish and are useful in providing variety to the diet or specialist feeding needs of some species. These include tubifex worms, mosquito larvae, bloodworms, water fleas (*Daphnia, Moina* and *Brachionus* spp.) along with brine shrimp (*Artemia salina*). Freeze-dried feeds will normally have one single ingredient, such as mosquito larvae, blood worms or similar. Sometimes the individual organisms can be distinguished or in other cases they will be sold as form pressed sheets or chunks. Since they contain a single organism they are not a complete diet – they are instead used to supplement other types of feed. Ideally using several different types of freeze-dried feeds will give the fish even more variation. Freeze dried feed is popular among aquarists who do not wish to store feed in their freezer and go through the process of thawing it before each feeding. Just as with manufactured feeds, it is very important to store freeze-dried products under optimal conditions and refrain from using old products.

Frozen Feeds

Commercial frozen feed can be either purchased or cultivated as large batches of live feed at home and frozen it into suitable serving pieces. Homemade feed preparations can naturally also be frozen to preserve their nutritional value. It is important that frozen feeds need to be thawed before serving them to the fish. It is excellent for stimulating fish that are suffering from loss of appetite to feed again. Brine shrimp, plankton, krill, and bloodworms are popular forms of frozen feed.

Granular Feed

Granules are like very small, hard flakes or tiny pellets. Currently only a limited variety of fish feed granules are available, usually engineered for the general nutritional needs of small community fish. These are designed to sink quickly to the bottom, for the benefit of bottom feeders.

Tablets

Tablets are just large flat pellets. Most tablets are of a sinking variety, but there are some that are engineered to adhere to the side of the aquarium to observe the fish feeding. Most of the sinking tablets are engineered to provide for the nutritional needs of scavengers and bottom feeders. These are designed to sink quickly **to** the bottom, for the benefit of bottom feeders.

Special Feed Types

The various types of micro particulate diets are categorized into three groups, micro-encapsulated diet, micro-bound diet and micro coated. Micro-encapsulating a solution, colloid or suspension of diet ingredients with a membrane is the micro encapsulation process. Micro-coated diets are prepared by coating micro-bound diet with some materials such as zein or cholesterol-lecithin.

Microdiets have been formulated to replace reliance on live feeds and weaning the fish larvae at their early stages onto microdiets. The chemical and physical properties play a crucial role with the behaviour of microdiet particle in the water and its attractability, leaching, ingestion and digestion.

Types of manufactured microdiets:

a. Microbound diets (MBD)
b. Micro-coated diets (MCD)
c. Micro-encapsulated diets (MED)
d. Marumerisation (MEM)

(a) MBD

Currently, the manufacturing process of MBD is the simplest and most commonly used method of preparation. It consists of dietary components held within a gelled matrix or binder. They do not have a capsule and it is suggested that this facilitates greater digestibility and increased attraction through greater nutrient leaching. Some commercial microdiets are manufactured using extrusion and then crushed and sieved to the required particle sizes. All the ingredients are ground, mixed with a binder such as gelatine, alginate, zein, carrgeenan and carboxymethyl – cellulose, activated by temperature or chemically and the dried (drum drying or spray drying), ground and sived to the required size.

(b) MCD

Micro coating method is based on coating or binding small MBD particles to reduce leaching. The coating layer is usually lipids or lipoproteins.

(c) MED

MED particles are made by using several different techniques. The particle usually has membrane or capsule wall, which separates dietary materials from the surrounding medium. The capsule wall helps maintain the integrity of the feed particle until it is consumed preventing leaching and degradation of the nutritional ingredients in the water. However, this attribute may restrict leaching of water – soluble dietary components and therefore reduce the larvae's attraction to the feed particles. The capsule wall is also thought to impair digestion of the feed particle. There are several methods for micro-encapsulation. These include chemical processes and mechanical processes. In chemical processes, the capsules are made within a liquid, usually stirred or agitated. The capsules are formed by i) spraying droplets of coating material on a core ingredients, ii) capsulating liquid droplets containing the nutritional ingredients by spraying into gas phase, iii) creating gel capsules by spraying droplets, containing the nutritional integredients and a binder, into liquid solution that activate the binder or by polymerization reaction at a solid/gas or iv) liquid interfaces. Protein cross – link (Yufer *et al.*, 1998) involves several stages of mixing and washing with organic solvents resulting in a very expensive diet process, as well as potentially toxic. These methods never resulted in good growth rates due to the larvae inability to digest and assimilate the particles as well as the high ratio of non-nutritional ingredient mainly the capsule, to the essential nutritients.

Another method, complex conversation, involves mixing and activating, using electrical charges, two liquid phases differing in their viscosity resulting in very small capsules. These capsules then bind to create a larger capsule containing hundreds or thousands of microcapsules.

(d) MEM

Mechanical encapsulation involves processes such as spray drying, fluidized bed drying, cold micro – extrusion marumerization (MEM) and particle – assisted rotational agglomeration. MEM is a two-step process of cold extrusion followed by marumerization (spheronization). The process has the capability of producing particles from 500 – 1,000µm and greater. Particle –assisted rotational agglomeration (PARA), which is a single step process capable of producing particles from 50 – 50µm that are lower density than particles produced by the MEM method due to the fact that the extrusion step is avoided. The method is based on a spinning disk (marumerizer). A wet mash of the ingredients is put into the marumerizer with or without inert beads. The rotation movement of the disk brakes down the mash into smaller spherical particles. The diameter of the particles depends on several factors including the disk rotation speed, the inert beads and the raw ingredients.

Hydrophilic amino acids leached the most from MBD, hydrophobic amino acids were found to leach from MED particles at a higher rate.

A diet particle needs to achieve a fine balance between leaching amino acids and other nutrients to act as feed attractant and digestibility of the particle to suit the undeveloped larvae digestive system. A particle that will be hard and leach resistant will also present a challenge to the larvae digestive system, whilst, a particle that will digest easily in the gut will also disintegrate relatively fast in the water.

Feeding Strategies

The stomach of fish is also generally adapted to the kind of feed they eat. Predatory fish generally have sac shaped stomachs that allow them to have enormous amounts of feed. Feed that is partially digested moves from the stomach into the intestine. Here it is digested further and the nutrients are absorbed into the body. The herbivores have an elongated intestine and their systems are more complicated than the carnivores. Feed that is not completely digested and absorbed leaves the body through the anal opening, together with other waste products produced by the metabolism.

Feeding Ornamental fishes

Giving the fish the right kind of feed at the right time and in the right amounts is crucial to their growth and development. In a closed system, the fish have no choice but to eat what they are provided with. If fish have to be full of energy and healthy need to be fed according to their particular needs.

- ☆ Each type of fish has a specific feeding requirement. Herbivorous fish require lots of fiber in their diet, while carnivores require feed that is rich in protein. Some predatory fish eat feed only after chasing the live feed.
- ☆ Do not make your fish obese. Yes, fish also gain unhealthy fat from over eating. Some fish like the catfish will eat any amount of feed. They become too big and lose health. Remember that fish in the aquarium do not expend energy looking for and chasing feed. They only move about within the confines of the tiny little aquarium.So,feed the fish only according to their nutritional needs.
- ☆ Overfeeding introduces a lot of unwanted toxins into the system. Mostly, fish are able to take in the feed they need within 5-10 minutes of their feed. Feed left in the aquarium after the first 10 minutes of feeding is not needed by the fish, and will collect in the aquarium as waste; decaying and releasing toxins.
- ☆ Supplying feed for the fishes can become a problem in a community or biotope aquarium at times. Hyperactive fish and fast swimmers will get the first feed in the aquarium. Small fish can be scared away by larger fish and newly introduced fish may be too shy to get to the feed.
- ☆ Ornamental fishes need to be given a variety of feeds. Feeding the fish with the same kind of feed day in and day out tends to decrease their appetite. Besides, they also need a variety of nutrients, which can be provided only by rotational feeding.
- ☆ We have to be careful when introducing live feeds into the aquarium. Many fish species love worms, insect larvae, etc and will stay much healthier when provided with live feed. Care must however be taken to ensure that these feed varieties do not carry infectious organisms, such as bacteria or other parasites. This is very difficult to ascertain, unless live feed is cultured in captivity.
- ☆ Feed for the fish has to encompass a large number of nutrients. All these together will make the fish healthy and make them strong to adapt to

changing conditions in the aquarium. The healthier the fish, the more resistance will they have to disease and infections.

Based on the feeding nature of ornamental fishes the ornamental fishes can be classified as:

1. **Top-feeders**: (Example : Mollies, Platys and Guppy) these fishes have a scoop like mouth which help them eating feed from the water surface. They prefer to feed on or near the water surface.
2. **Midwater-feeders**: (Example: Danios and Minnows) mouth of these fishes are located at the very tip of their snouts to gather feed as it falls through the water.
3. **Bottom-feeders**: (Example: Catfish and Corydoras) these fishes have mouth beneath their snouts and they feed from floor or bottom surface.

Feeding Method

1. **Manual Feeding**: This is most common method for feeding fish. If there is a mechanical filter and feed floating on the surface is drawn into the filter, the portions of feed should be put below the water surface and away from the filter. Feed will not accumulate in the filter.
2. **Automatic Feeder**: Manual feeding is better, but may not be suitable for people with very busy schedule or with unplanned holidays. For such people it is worth considering a automatic feeder. It is also a good alternative to vacation feed. The automatic dispenser releases a definite amount of feed at regular intervals.

Diets Formulated for Colour Enhancement

Fish that are colored in nature often acquire faded coloration under intensive culture conditions. Experiments adding top-coated algae to the diets of ornamental fish have resulted in color enhancement. Freshwater red velvet swordtails *Xiphophorus helleri*, rainbow fish *Pseudomugil furcatus*, and topaz cichlids *Cichlasoma myrnae* became significantly more intensely colored when fed a diet containing 1.5-2.0 per cent of a carotenoid-rich strain of *Spirulina platensis* and 1.0 per cent of a specially grown *Haematococcus pluvialis* for 3 weeks. Though color enhancement was apparent after only a week, when the fish consumed these doses of algae, lower doses (0.5 per cent and 0.4 per cent, respectively) were not shown to give coloration. Color enhancement appeared to occur via natural carotenoid receptors. Thus, color intensity diminished when fish were stressed.

Colour enhancement was environment-sensitive. Topaz color in cichlids developed only after the aquaria were divided into territories, and rosy barb color intensified when floating substrate was present. It is concluded that ornamental fishes are good models for color enhancement through diet and that this enhancement may be achieved using products made by marine biotechnology companies.

Since they use natural algae that mimic the absorption of carotenoid that occurs in the wild, they should be more acceptable to consumers who may have concerns about the use of chemicals or hormones to enhance color. A "cocktail" of algae supplying natural stereoisomers of ß-carotene, zeaxanthin, lutein, canthaxanthin, and astaxanthin is used because fish sometimes seem to metabolize carotenoids before depositing them onto natural receptors in the skin in a species dependent way This was observed in preliminary results whereby the blue-green fluorescent colors in discus fish *Symphysodon sp.* is enhanced by feeding sources of ß-carotene and red colors is enhanced by feeding sources of canthaxanthin and astaxanthin. However, the red color of red swordtails and tinfoil barbs *Barbus schwanenfeldi* was enhanced by ß carotene sources.

Feed Preparation

The methods begin with the formation of a dough-like mixture of ingredients. Ingredients can be obtained from locally available sources. The dough is started with blends of dry ingredients which are finely ground and mixed. The dough is then kneaded and water is added to produce the desired consistency depending upon the fish species which is going to be fed. Pelleting or rolling converts the dough into pellets or flakes, respectively. The amount of water, pressure, friction, and heat greatly affects pellet and flake quality. For example, excess water in the mixture results in a soft pellet. If the feed has too little moisture the pellet will crumble. Proteins and especially vitamins are seriously affected by high temperatures. Therefore, avoid storing diet ingredients at temperatures at or above 70° C (158° F) and do not prepare dry feeds with water at temperature higher than 92° C (198° F).

Feed Equipments and Storage

Making own fish feed requires few specialized accessories. They are used primarily for chopping, weighing, measuring ingredients, and for blending, steaming, cooking and drying the feed. Most of the utensils needed will already be in the laboratory or kitchen. Multipurpose kitchen hand graters, a knives for cutting, slicing, and peeling can be used. A couple of plastic cutting boards protect the counter and facilitate handling the raw ingredients. Heat resistant rubber spatulas, wooden and slotted spoons, long-handled forks, and tongs are very good for handling and mixing ingredients. A basic mortar and pestle, electric blender, feed processor or coffee grinder are very useful to chop or puree ingredients; use grinder sieves and mince die plates to produce the smallest particle size possible. A feed mill and strainer such as a flour sifter help discard coarse material and obtain fine feed particles. For weighing and measuring ingredients, dry and liquid measuring cups and spoons, and a feed weighing scale is required. Other utensils include plastic bowls for weighing and mixing ingredients. A wide mouthed saucepan is good for heating gelatins and cooking raw feeds such as vegetables and starches. The ingredients and blends may be cooked in a small electric or gas burner. Ingredients may be mixed by hand using a rotary beater or mixer, however, an electric mixer or feed processor is more efficient. After mixing, a dough is formed which can be made into different shapes.

An extruder will extrude the dough into noodles or "spaghetti" of different diameters. As the noodles emerge from the outside surface of the die, they can be cut off with a knife to the desired length or crumbled by hand, thus making pellets. A potato dicer also serves to extrude the dough into noodles of the same size. The pellets or thin sheets can be placed on a cookie sheet and dried in a household oven on low heat or in a forced-air oven. A small feed dehydrator also performs the task quite well. To add extra oil and/or pigments to pellets, a hand-held oil atomizer or sprayer can is useful. To separate pellets into different sizes, a set of sieves (*e.g.*, 0.5, 0.8, 1.0, 2.0 and 3.0 mm) is required.

Freezer bags serve to store the prepared feeds, and using a bag vacuum sealer will greatly extend the shelf-life of both ingredients and the feed. The feed can be stored double bagged in the freezer but should be discarded after 6 months.

A formulated pellet diet, especially used for experimental purposes, should be analyzed for nutrient content (proximate analysis for crude protein, energy, moisture, etc.). In addition, before intending to make own fish feeds with unfamiliar ingredients, it should be analyzed prior to their use.

Sample Formulations and Recipes

There are numerous recipes for making fish feeds. Purified and semi-purified diets are used primarily in experimental formulations to study the effects of nutrients, such as the amount or type of protein, may have on the health and growth of fish. One simple formulation, which is used traditionally to feed ornamental fish in ponds, consists of a mixture of 30 per cent ground and processed ragi flour, 30 per cent of fish meal and 20 per cent soybean flour. By weight, approximately 2-3 per cent of fish oil, and a 0.3 per cent vitamin and a 1 per cent mineral premix are added to the mixture. This mixture is blended with water and can be formed into dough balls of different sizes. Care should be taken so as not to pollute water by over feeding. The dough can be steam cooked to give stability and then fed to ornamental fishes. The vitamin and mineral mixture should be added after steam cooking to avoid loss of nutrients.

References

Boonyaratpalin, M. and R. T. Lovell. 1977. Diet preparation for aquarium fishes. *Aquaculture* 12: 53-62.

Chong ASC, Ishak SD, Osman Z. and Hashim R. 2004. Effect of dietary protein level on the reproductive performance of female swordtails *Xiphophorus helleri* (Poeciliidae). *Aquaculture*; 234(1-4): 381-392.

De Silva, S. S. and T. A. Anderson. 1995. Fish Nutrition in Aquaculture. Chapman and Hall. London, UK.

Pannevis MC, Earle KE. 1994. Maintenance energy requirement of five popular species of ornamental fish. *J Nutr*; 124(Suppl): 2616-2618.

Pannevis, M. C. and K. E. Earle. 1994. Nutrition of Ornamental Fish: Water Soluble Vitamin Leaching and Growth of *Paracheirodon innesi*. *Journal of Nutrition* 124: 2633S-2635S.

Sales J, Janssens GPJ. Nutrient requirements of ornamental fish. 2003. *Aquat Living Resour*; 16(6): 533-540.

Tacon, A.G.J. 1987. The nutrition and feeding of farmed fish and shrimp - A training manual 1: The essential nutrients. Food and Agriculture Organization.

Tamaru CS, Ako H, Paguirigan R. Essential fatty acid profiles of maturation feeds used in freshwater ornamental fish culture. *Hydrobiologia* 1997; 358(1):265-268.

Wallat GK, Lazur AM, Chapman FA.2005. Carotenoids of different types and concentrations in commercial formulated fish diets affect color and its development in the skin of the red oranda variety of goldfish. *N Am J Aquac;* 67(1):42-51.

Chapter 12

Feeding Receptors in Ornamental Fish Feeding

Cheryl Antony

Institute of Fisheries Technology
Tamil Nadu Fisheries University,
Chennai – 51

The aim of the ornamental fish farmer should be the efficient conversion of healthy, live fish, with the minimum of feed wastage. Therefore any way of increasing the acceptability or palatability of fish diet would be advantageous. Fish vary considerably in their willingness to accept compounded diet. Fishes are either visual feeders or some use their chemical senses, smell and taste, in the detection of feed.

Factors Affecting Feeding

The efficiency of utilisation of feed particles is affected by many external and internal factors. Primarily the searching, identification and ingestion processes are influenced by physical and chemical factors including colour, shape, size, movement and olfactory stimuli at molecular level.

To be acceptable, a diet must satisfy the following criteria:

1. Appearance, size, shape and colour.
2. Smell: which in the case of aquatic animals, should be termed long-range chemical attraction, since such animals can use both smell and taste to detect feed at a distance.
3. Feel of the material : hard or soft, most or dry, rough or smooth?
4. Taste : taste buds in the mouth, monitor the taste of the material.

Which are the most important features of a feed which depends on whether the particular fish is a visual or chemosensory feeder? The chemical activators can be divided into three groups. *Attractants* guide the fish towards the feed, *incitants* invoke biting and tasting, and *feeding stimulants* induce the fish to swallow the feed.

Feed Identification and Ingestion

There are several steps in the feeding process of fish larvae in the process of finding and ingesting feed particles.

1. General and not specific reaction, initiation of search movements involving chemical and electrical stimuli.
2. Identification of the feed particle, involving chemical stimuli
3. Identification of the feed particle by location chemical stimuli
4. Close identification of the feed particle location involving chemical and visual stimuli
5. Tasting and/or actual feeding requiring chemical stimuli (taste buds)

Identifying Feed Stimulants

It is the nature of the gustatory feeding stimulants in the feeds. The test diet was based on casein and the unflavoured diet was unacceptable to most fish. However, when a mixture of chemicals, based on an analysis of squid mantle tissue (synthetic squid mixture) was added at a level of 1–2 per cent, the fish readily ate the diet. The feeding stimulant could then be identified by adding various components of the complete mixture to the casein diet and measuring the quantity of test diet eaten.

Another possible factor in the "choice" of feeding stimulant is the feeding habits of the indigenous fish under natural conditions. Free L-amino acids, the feeding stimulants for fishes are present in all animal tissues, both vertebrate and invertebrate. Different amino acid mixtures acted as feeding stimulants for the various species, and in all cases the corresponding D-amino acids were ineffective.

Various substances, such as free amino acids, nucleotides, nucleosides and ammonium bases are potent inducers of feeding behaviour. Inosine and inosine 5'-monophosphate are also the specific feeding stimulants. Synergistic relationship is observed between several amino acids (glycine, alanine and arginine) and betaine which when combined produced stronger effects.

The crustacean feed sources all contain glycine, betaine and free L-amino acids, the feeding stimulants. It is generally considered that teleost fish contain a little or no glycine while invertebrates and elasmobranchs are rich in this chemical. A much wider range of fish species will have to be investigated before any definite conclusions can be drawn, and before any predictions can be made regarding the chemical nature of the feeding stimulant of an untested species. However, fish -eating species most probably have amino acid mixtures or inosine 5'-monophosphate as feeding stimulates and these eating invertebrates, glycine, betaine plus amino acids.

Amino Acids vs. Hydrolysates as Feed Attractants

Amino Acids

- Only the L-isomers have been found to be active as feed attractants.
- Various combinations of amino acids have been found to have a positive effect on various fish species.
- Synergistic effects were associated with many combinations of amino acids and other substances such as ammonium salts.
- Increasing the concentration of amino acids (when added to the water)was found to have positive effects on feeding range from 10^{-8} M to 10^{-2} M.

Hydrolysates

- Hydrolysates act as feed attractants as they contain digested protein components such as free amino acids and peptides.
- Concentrations of extracts and/or hydrolysates made from aquatic animals have a positive effect on feeding range from 10^{-2} to 10^{-10} g/l (when added to water).
- The protein fraction weight between 1000 and 10,000 Dalton is found to have a positive effect on feeding.

Presentation of Feed Attractants to Fish Larvae

The addition of attractants directly into the water uses large amount of theses substances, but maintains a constant concentration.

Coating the diet particle in unknown leaching rates but can contribute to higher palatability and more specifically identifies particles as feed.

By incorporating into the diet, as part of the protein source also results in an unknown leaching rate (depending on the type of microdiet) however only a low amount of attractants are required, part of the protein source in the diet is replaced and digestion and assimilation is improved.

Feed Additives

Feed additives are substances which are added in trace amounts to a diet or feed ingredient either a) to preserve its nutritional characteristics prior to feeding (*i.e.* antioxidants and mould inhibitors), b) to facilitate ingredient dispersion or feed pelleting (*i.e.* emulsifiers, stabilisers and binders), c) to facilitate growth (*i.e.* growth promotants, including antibiotics and hormones), d) to facilitate feed ingestion and consumer acceptance of the product (*i.e.* feeding stimulants and feed colourants), or e) to supply essential nutrients in purified form (*i.e.* vitamins, minerals, amino acids, cholesterol and phospholipids).

Characteristics of Feeding Stimulants

Maximum benefit from feeding can only be achieved if the feed provided is ingested. An understanding of the feeding behaviour of the ornamental fish is, therefore, essential. The diet presented must have the correct appearance (ie. size, shape and colour), texture (ie. hard, soft, moist, dry, rough or smooth), density (buoyancy) and attractiveness (ie. smell or taste) to elicit an optimal feeding response. However, the relative importance of these individual factors will depend on whether the fish species is mainly a visual feeder or a chemosensory feeder. Fishes held in captivity generally rely on sight to locate their feed, they also rely on chemoreceptors located in the mouth or externally on appendages such as lips, barbels and fins; the feed being carefully 'sensed' before ingestion. The use of dietary feeding stimulants is therefore essential to elicit an acceptable and rapid feeding response. The practical importance of feed attractants and diet palatability is particularly critical during the weaning of fish larvae from a live to a non-living diet. Similarly, as attempts are made to replace the fishmeal component of practical fish feeds with unconventional protein sources of a new nature, the problem of diet texture and palatability will become even greater. In addition, by using feeding stimulants and improving feed palatability, the period of time the feed remains in the water can be reduced, thus minimizing nutrient leaching.

Two types of feeding stimulants may be considered for use within aquaculture feeds; natural ingredient sources which exhibit attractant or feeding stimulant properties or the use of the purified or synthetic chemical derivatives which are responsible for the attractant property of natural ingredient sources. For example, feed ingredients which have been found to impart specific attractant properties include squid meal, mussel flesh, shrimp meal and waste, short-necked clam flesh, marine polychaete worms, blood worms, certain terrestrial oligochaete worms, marine fish oils, fish meal, fish solubles, fish protein hydrolysates and soybean protein hydrolysates.

Purified or synthetic substances which have been found to act as dietary feeding stimulants include mixtures of L-amino acids (particularly amino acid mixtures including glycine, alanine, proline and histidine: mixtures of L-amino acids and the quaternary amine glycine betaine, the nucleosides inosine and inosine-5-monophosphate, the nucleotide uridine-5-monophosphate and trimethyl ammonium hydrochloride. Good dietary sources of betaine and soluble nucleotide bases include mussels, polychaetes, squid, shrimp waste and shrimp blanch water, fish products, polychaetes, respectively.

Feeding Response Factors

Feeding behaviour may be controlled by several chemical factors: attractants, in citants (bitingfactors) and stimulants (swallowing factors). Visual feeders are, by definition, attracted to their feed by sight and long-range chemical attraction is not involved. The chemical sensès are used t o detect feed, and mixtures of chemicals serve as the attractant. Inosine and inosine 5-monophosphate are specific feeding stimulants. All fish are primarily sight-feeders, while some fishes, detects its feed a t a distance, by means of smell or taste, and glycine betaine in this case acts as attractant and feeding stimulant.Amino-acids are required in addition to glycine betaine for juvenile fishes.

The chemical nature of the feeding stimulant for sight - feeding fishes, showed that the feeding stimulant activity resided in the L-amino-acid fraction, the non-amino-acid components being completely inactive. Among the amino-acids it shows that the mixture of neutral L-amino-acids was highly active, while the corresponding D- aminoacids were inactive. However, the two major neutral aminoacids, proline and glycine, are together essentially inactive and also the neutral amino-acids, indicating synergistic effects

Chemo-Attractive and Feeding Stimulatory Effect of Dietary Nucleotides

The low-molecular weight fraction of squid and hypothesized nucleotide (AMP) and nucleoside (inosine) components are the main chemo-attractants for aquatic animals. The presence of chemoreceptors on the lips of the fishes that responded to nucleotides (AMP, IMP, UMP and ADP) have been analyzed by electrophysiological methods. These experiments resulted in an important discovery of the chemoattractive effect of dietary nucleotides on fish. On testing 47 nucleosides and nucleotides it was identified that inosine and IMP are the most potent gustatory feeding stimulants for most bottom dwellers based on feeding behavior of fish fed experimental diets. The behavioral or gustatory responses of fishes to exogenous nucleotides may be species specific. Person-Dietary inosine enhanced growth of fish larvae. Larvae also fed a diet supplemented with both betaine and inosine showed significantly higher growth than larvae fed diets supplemented only with betaine or inosine and a reduced amount of betaine.

To date, it has been reported that only IMP, but not inosine has stimulatory effects on feeding of fish species. Dietary supplementation of IMP (2800 mg kg_1) enhanced feed intake by 46 per cent compared to the non-supplemented soybean meal-based diet. IMP may serve as a primary candidate for feed attractant research to further explore complete replacement of fish meal in aquafeeds.

Betaine, L-alanine, L-glutamic acid, L - arginine, glycine and inosine are know as dietary feeding attractants for many fish species. However, these additives also have very important physiological functions within the animal body. For example, Betaine given singly or fixed with other attractants has been found to have a positive effect on fish growth and survival rate. Betaine is also a very important substance for methyl donation and osmoregulation during the transfer of salmon and trout from freshwater to seawater. Similarly, L- alanine and L- glutamic acid can readily enter the citric acid and be used for energy supply in fish and fish larvae. Moreover, L - arginine is an essential amino acid for fish, glycine is an important constituent of collagen and elastin and inosine is a nucleoside and has vitamin activity in many fish species.

The use of dietary feeding attractants within compound aquafeeds has received considerable attention in recent years. The rationale behind their use has been to improve dietary feed intake and at the same time by promoting quicker feed intake, minimizing the time the feed remains in water and thereby the leaching of water soluble nutrients and at the same time providing additional nutrients for protein and energy metabolism. It follows therefore that if aquafeed are ingested with minimum wastage feed efficiency will be improved and water pollution will be minimized.

Betaine

Betaine (glycine betaine, trimethlyglycine) is a highly water soluble and therefore diffusible compound which has the ability to stimulate the olfactory bulb of fish. It is found in high quantities within marine invertebrates (Meyers, 1987), micro organisms and some plants, and as such constitutes an important part of the natural diet of marine carnivorous fish and crustacean species.

L-amino Acids

Although the use of conventional and unconventional plant protein sources as dietary fishmeal replacers within aquafeeds is nutritionally feasible, the resultant feeds are generally much less palatable to the farmed species. It follows therefore that to overcome these difficulties (at least in terms of palatability) that these diets be supplemented with dietary feeding attractants and stimulants.

In this respect, like betaine, free amino acids are also high water soluble and easily diffused in water. In particular, L-alanine, L-glutamic acid, L – arginine and glycine have been reported to have dietary attractant properties; alanine, glutamic acid and glycine being non essential amino acids, and L-arginine being an essential dietary amino acid for fish.

Betaine and certain free amino acids give the opportunity to fish nutritionists in preparing more palatable feed for fish during the different growth stages. As a mixture of attractive substances, they can usually beneficially affect both feed consumption and fish growth. It is generally accepted that the use of dietary feeding attractants and stimulants within aquafeeds will increase in the future as the industry is forced to use more economical alternative dietary protein sources as fishmeal replacers.

Feeding Patterns/Mode in Ornamental Fishes

The feeding of fish and their nutrition is one of the most important factors in keeping them healthy. Since fish are very diverse in their habitats, there is a large diversity in their eating patterns too. Some fish, for instance, are bottom feeders, while others are mid-water feeders or surface feeders. Most of the fish that have adapted to aquarium life can be trained and habituated to eat different kinds of feed.

- ☆ The first step is that the fish has to be able to recognize the feed. Even the most nutritious of feed goes to waste if the fish does not understand that it is to be eaten.
- ☆ Both instinct as well as training affects this recognition. Hunger is not the only thing that leads a fish to feed. Various visual and chemical clues point out the feed to the fish.
- ☆ Once the fish has located the feed, it may taste it before it accepts the feed. Some fish may take in the feed, and regurgitate it if they feel that it is not acceptable.

- ☆ Predatory fish have a different kind of feeding pattern. Generalizations about the sensory characteristics of diverse kinds of fish will therefore not be accurate or appropriate, but here are some general traits in feed that can attract fish.

Flavor and Taste

This characteristic of feed is especially important in case of bottom feeders. Smell can be detected by the specific anatomical receptors in fish, but flavor has to be dissolved in water for the fish to locate it. Some fish have receptors in their mouths, or on the head or lips. Some even have taste receptors on their skin. These receptors carry messages to the brain and tell the fish to swim towards the feed. Some kinds of feed can strongly stimulate fish to feed by their flavor.

Sound

Through water, sound travels about four times faster than it does through air. So, a fish can actually "hear" sound through the vibrations that take place in water. By picking up these vibrations in water, the fish become aware of the feeding frenzies that cause many fish to conglomerate when the feeding begins. Also, there are fish that are so used to a routine in their feeding that they start grouping when they hear sounds that normally precede feeding.

Smell

The sense of smell is highly developed in fish. In nature, fish needs to be able to identify their feed and also their mates through the sense of smell. So, many fish species have nostrils that help them to identify the various things they come across. These sensors thus help the fish to feed on their feed.

Colour and Buoyancy of Feed

Some fish that are used to feeding on floating feed may not take feed that has sunk to the bottom. Similarly, bottom feeders rarely come to the top of the aquarium to eat feed. A majority of the tropical variety of fish species are however not very choosy when it comes to the buoyancy of feed.

A complex cross linked protein walled microcapsule by encapsulating lipid wall capsules containing water soluble nutrients along with other dietary nutrients into a cross linked protein membrane. The advantage of this diet is its ability to release low molecular weight attractants (*i.e.* Free amino acids, nucleotides) from the protein wall while other essential nutrients such as water soluble vitamins are retained in the lipid wall capsules encapsulated in the diet.

Classification of Taste Substances and Feeding Mechanisms Employed by Fish

Chemical substances can be divided into several categories or types, according to their effects on the feeding behaviour of fish. There are distinctions between the substances that have effects on the oral (gustatory) and extraoral (olfactory) taste systems. These systems play different roles in fish feeding behaviour and differ in their functional characteristics. The presence or absence of these compounds in the diet determines whether a feed item is grasped or ignored, is eaten or rejected and to some extent, how much of the feed is consumed.

Chemo- attractants effecting feeding behaviour in fish are classified as follows:

- 'Incitants' are substances that induce capture of the feed item via the extraoral taste system. There are a number of different behaviors that can presumably be evoked by incitants like: suction, grasping, snapping, biting, tearing or pinching.
- 'Suppressants' are substances that decrease the rate of grasping feed items. Like incitants, suppressants are mediated by the extra oral gustatory system and control the rate of grasping feed items.
- 'Stimulants' are substances characterized by high ingestion rate. This behaviour is evoked by the oral taste system. Stimulants promote fish feeding. Usually, fish swallow the feed items containing stimulants at first capture. This behaviour is mediated by the oral gustatory system.
- 'Deterrents' are substances that make the fish abandon feed intake and evoke feed rejection. Deterrents are characterized by high rejection and low ingestion of feed items that are caught. Often, fish feeding motivation is decreased for a short period after a fish has tested a feed item that contains a deterrent. Usually, the retention time of feed items that contain a deterrent substance is short and the fish does not attempt re-catch the feed item. This behavior is mediated by the oral gustatory system.
- 'Enhancers' or 'potentiators' are substances that, although not feeding stimulants, accentuate the flavor of the feed and cause fish to increase their consumption of flavoured feed. In some cases, however, stimulants may be ineffective as feeding enhancers. This behaviour is evoked by the oral taste system.

A single substance might have different effects on the taste behaviour of the same fish and can be both incitant and deterrent, or incitant and stimulant. A number of substances do not induce any effects on fish behaviour either via the extraoral or the oral systems and are classed as 'indifferent' substances.

In the location and identification of feed, fish may rely on one a combination of the following sense stimuli.

Visual Detection

Visual identification, eye sight can play an important role in locating and identification of feed. Visual location and identification decreases at night fall or when water body is murky, as is the case with water that has a high level of suspended solids (dark brown water).

Electro Detection

Electro-sensory detection in fish is achieved with the use of small electro pulses which are emitted by the fish itself, these pulses are then interoperated as they reach the fish again, after reflection by feed.

Mechano-Reception

Fish use stimulus cues like sound and turbulence to locate feed. This stimulus is registered by the mechano-receptors located on the lateral line.

Chemo-Reception

Fish located feedstuff by following chemical signals produced by its respective prey of feed, this is know as olfactory response. The final consumption of the feed item is determined by the gustatory response of the fish towards the feed item, this determines the palatability of that the feed item.

Various types of fish behaviour are mediated through olfaction. Some fish employ the sense of smell to find feed; odours may be released by freshly killed, injured or already decaying organisms or the odour of a prey can be significant to a predator.

The gustatory (taste) sensory system provides the final evaluation in the feeding process. Gustatory responses to amino acids are highly stereo – specific. The L-isomers of amino acids have, for most species studied, been shown to be more stimulatory than its D-isomers. Unsubstituted, L- amino acids are usually the most efficient compounds for the gustatory system of fish.

General findings so far indicate that, i) only a - amino acids are highly stimulatory, ii) L-isomers are always more stimulatory than D- isomers and iii) the stimulatory effectiveness is not directly related to the essential amino acids. Neutral amino acids containing two or fewer carbon atoms, and having un branched and uncharged side chains, are highly stimulatory. Acidic amino acids are poor gustatory stimuli, and basics amino acids are quite variable, depending upon the fish species. A synergistic effect was observed for mixtures of some amino acids in the gustatory systems of several fish species.

Feeding Factors

In higher animals, feeding behaviour may be controlled by several chemical factors: attractants, incitants (biting factors) and stimulants (swallowing factors). Visual feeders are, by definition, attracted to their feed by sight and long-range chemical attraction is not involved.

Essentially all the feeding stimulant activity resided in the L- amino acid fraction, the non-amino acid components being completely inactive. However, the two major neutral amino acids, proline and glycine, were together essentially inactive, as were the remaining neutral amino acids, indicating synergistic effects.

Feeding in fish, as in other groups of animals, is an important function of life, and is a result of processes that are associated with searching, aiming, accepting, seizing, oral processing and evaluation of the quality of feed objects. Swallowing, digestion, absorption and assimilation follow these processes. Directly related to the satisfaction of energy requirements is responsible for the rate of growth, maturation and fecundity of fish, their social status, migratory activity, life strategy and their resistance to the impact of unfavourable factors.

The consummatory phase of the feeding behaviour starts with the awareness of a feed object and terminates with either swallowing or rejection. The palatability (taste) of a feed object for the fish involves predominantly gustatory and some aesthetic qualities. Various mechanosensory organs, and what is termed as a common chemical sense, can mediate these latter qualities. In many instances, a fish rejects a feed item after it has been taken into the mouth cavity.

Such behaviour demonstrates that an object that is selected as a feed item, based on the information given by any of the sensory systems available is not sufficient to guarantee its suitability as feed for the fish. The sensory systems in question are olfaction, vision, acoustic, lateral line organ, extraoral gustatory subsystem or electroreception. This consideration indicates the important control functions of receptors located within the fish's mouth cavity in evaluating the specific feed items.

Sensory System in Feeding

The gustatory sensory system provides the final evaluation in the feeding process.

However, their sensory systems also participate in the consummatory phase in fish. The sense of touch or mechanoreception is functionally and structurally connected to the gustatory system.

The Gustatory System in Fish

The peripheral organs belonging to the gustatory system are the taste buds. Taste buds constitute the structural basis of the gustatory system in all fish. In fish, the taste buds are situated not only within the oral cavity pharynx, oesophagus and gills, but may also occur on the lips, barbels and fins and over the entire body surface in many species.

The abundance of taste buds is another peculiarity of the fish gustatory system. Taste buds are more numerous in fish than in any other animal.

Feeding Behaviours Mediated by the Gustatory System

Various sensory organs can mediate the different feeding behaviours found in fish and a particular behaviours pattern can be evoked via several sensory systems. For example, the snapping movement with the jaws can be released by stimulation of the lateral line organ or electroreceptor, by visual auditory or olfactory stimuli or by taste or common chemical sensory stimuli.

Usually, the gustatory systems participate in the final phases of the series of feeding behaviours connected with feed search, which normally ends with consumption. It seems established that stimulation of the external taste system may mediate grasping biting, snapping or scraping behaviours, especially in the bottom feeders. Feed objects containing aversive substances cause fish to neglect the object by moving forward.

When a feed object is in the mouth, it is subject to final sensory judgments. As a result of these oral rests, the feed item can then be rejected or swallowed. The time for which the object is kept in the mouth before swallowing or rejection is called the retention time. During the retention time, the fish detect and recognize taste substances, assess the palatability of the feed object and perform the decision to swallow or in spit it out.

Classical Taste Substances

Sucrose

Sucrose is usually an indifferent taste substance for fish or evokes a positive oral taste response. Sucrose is palatable for many herbivorous fish like carp is palatable for omnivorous fish in which algae is the main component of the diet.

Free amino acids have been found to be highly efficient incitants or stimulants for various freshwater and marine fish. The spectra of stimulatory free amino acids are highly species – specific. Amino acids L. alanine, L- cysteine and L- serine act most frequently as a stimulant.

The amino acids are indifferent taste substances for fish more often than they are stimulants. In particular, L-norvaline, L – leucine, L- lisoleucine, L- asparagin, L-lysine, L-tryptophan, L-glutamine and glycine are most indifferent taste substances for fish.

Betaine

Betaine (glycine betaine, trimethylglycine) is widely distributed in fish feed organisms. The gustatory system of many fish species is highly sensitive to this substance. Betaine has important synergistic properties with amino acids or other substances. Mixtures containing betaine and amino acids were 9-16 times as effective as any of the other components used alone.

Sugars and other Hydrocarbons

Sugars can be stimulants, deterrents or indifferent stimuli for fish. Four sugars *viz.* D-fructose, sucrose, D-ribose and D-glucose, were palatable.

Sucrose is a stimulant for different poecilids species belonging to the family Pocilidae – the black molly and the platy.

Four sugars (arabinose, lactose, ribose and maltose) normally evoke negative taste responses in black molly; however, none of the sugars tested were deterrents for platy.

Other Types of Substances

Dimethyl - β- propiotethin (3-dimethyl – 3 – thiopropanol; DMTP) and dimethyl – thioproprionic acid (2 – carboxy – ethyl dimethyl sulphonimum bromide : DMPT) are very effective.

Thresholds

The minimal concentrations needed to give a positive or negative response of a fish towards giving a taste stimulus is an important parameter when it come to flavouring feed. Taste stimulants are similar within a fish family.

Relevance of Attractans to Aquaculture

The farming and husbandry of fish demand the adoption of intensively managed production systems, where feed management practices play an important role and requires serious research efforts for the generation of appropriate technology.

Supplementation of artificial, dry diets with attractants (feeding stimulants) can increase acceptability and consequently the consumption of low palatability diets by fry and fingerlings, which can increase growth rate and productivity.

Food flavor is mediated chemically by substance inherent to the food, and it is affected by the chemo- sensitivity of the species. Therefore, mixtures of attractants are more effective than individual component. L-amino acids are known to mediate stimulatory responses in fish, particularly the neutral amino acids, organic, animated bases and nucleotides are also involved in the stimulatory response of fish.

In aquaculture supplementation of artificial, dry diets with attractants (feeding stimulants) can increase acceptability and consequently the consumption of low palatability diets by fry and fingerlings. This practice can also reduce feeding time and feed waste, white improving water quality and environmental safety. Chemo attractants are species specify and thus species have their own range of chemo attractants. There is room for further research within the range of species related amino acids as chemo attractants.

The use of dietary feeding attractants within compound aquafeeds has received considerable attention in recent years. The rationale behind their use has been to improve dietary food intake and at the same time by promoting quicker food intake, minimizing the time the feed remains in water and thereby the leaching of water soluble nutrients and at the same time providing additional nutrients for protein and energy metabolism. It follows therefore that if aquafeed are ingested with minimum wastage and better feed wastage and water pollution will be minimized.

To develop microdiets (MD) to replace live feed, both rotifers and *Artemia*, as complete or partial replacements for fish larvae the efficiency of utilization of feed particles is affected by many external and internal factors. Primarily, the searching, identification and ingestion processes are influenced by physical and chemical factors including colour, shape, size, movement and olfactory stimuli at a molecular level.

Substances secreted by live food organisms that act to stimulate a feeding response belong to a group of chemicals known as 'feed attractants' and some have been specifically identified for larvae these physical and chemical factors affect the palette and influence the ingestion process, which is the precursor to the digestion process. Digestion involves secretion of enzymes, peristaltic movements and after larvae metamorphosis, acid and bile salt secretions. The assimilation and absorption process begins after the food particle is digested and broken down into more simple molecules that can pass across the gut lining. This is further facilitated by the development of brush border and microvilli as well as protein transporters and other transport mechanisms.

Various substances, such as free amino acids, nucleotides, nucleosides and ammonium bases, are potent inducer of feeding behavior in marine and freshwater fish larvae. Synergistic relationship was reported between several amino acids (glycine, alanine, arginine) and the ammonium salt betaine, which when combined produced a stronger effect.

Studies in Ornamental Fishes

The number of amino acids that evoke the same taste response are L-alanine, L-cystcine, L-phenylalaine, L- serine, L- theronine and L- threonine and L-tyrosine in black molly and platy.

There are examples of similarities in taste responses in closely related fish species. This is more evident among classical taste substances than among free amino acid. Sucrose was a stimulant for three Poeciliidate species tested. Citric acid was a deterrent for crucian carp with wild goldfish.

The palatability of numerous substances was different in common carp, crucian carp and wild goldfish that inhabit the same environment.

The difference in taste preferences are confirmed by the correlation analysis of amino acid palatability. Moreover, correlation between taste responses to free amino acids was negative in the pair, guppy – platy.

These data clearly demonstrate an important and undoubtedly leading role of gustatory reception in the feeding selectivity in fish and in their capacity to consume appropriate feed items that are specific to them.

The fish rejected flavoured pellets significantly more often than the blank pellets. Many of the fish preferred pellets flavoured by L-cysteine and showed an indifferent response for pellets containing L-glutamic acid. There were several specimens with a relatively low taste preference for L-cysteine and a very high taste preference for L-glutamic acid.

Specificity between Sexes

A study of sex variability in taste preferences of fish was performed on adult guppy. Citric acid, sucrose, calcium chloride, glycine, L-glutamic acid and L-histidine were used as taste substances. It was found that the males and females of guppy have similar taste preference for the substances tasted. Sucrose, glycine and L-glutamic acid were stimulants, calcium chloride and L- histidine had an indifferent taste for fish.

The results indicate that there are no differences between male and female guppy in taste preferences. It was shown that females and males use the same groups of feed objects, which include green algae and diatoms, benthos, detritus and larvae of insects, as feed. The duration of the taste response was shorter among males than among females.

Genetics of Taste Preferences

A study was performed by comparing taste preferences in wild female goldfish, male common carp and hybrids between these two cyprinid species. There were two main reasons to use these species – the palatability for some substances is different in these two species and goldfish and common carp readily yield hybrids. All specimens used for these experiments were reared on chironomid larvae and were maintained in water with similar temperature, oxygen supply and light cycle.

It was found that the taste preference for the classical taste substances were similar in carp and hybrids. However, the rank according to the preference of these substances was significantly different between hybrids and wild goldfish.

Based on a comparison of taste preferences in herbivorous and carnivorous fish, it was suggested that these two groups of fish have specific amino acids that stimulate feeding.

Internal Factors Effecting Taste Preferences

Feeding Experience

It is well known that fish release strong and obvious behavioural responses to feeding signals emanating from common feed objects. This has been shown for feed stimuli mediated by olfaction and vision. However, when it comes to taste preferences, the dependence upon feeding experience has been shown to be less prominent.

Conditioning with a long term rearing on a selected feed leads to only a small shift in taste preferences and only a slight increase in palatability for the substances in the feed on which they were reared. In conclusion, the results from these experiments provide evidence that taste preferences show low plasticity in fish. In other words, taste preference does not depend upon the fish diet. The fish diet might be reflected in taste preferences of fish to feed extract, which have a more complicated composition. The results showed that taste preference in fish is under strict genetic control.

Feeding Motivation

The following motivation strongly influences taste preferences. It was found that starvation during 17- 18 hrs caused a widening of the extroral taste.

Environmental Factors Effecting Taste Preferences

Temperature

The ambient water temperature is one of the most potent abiotic variables affecting vital function in exothermic animals like fish which also has an effect on taste preferences in fish.

The difference in water temperature between the 'warm' and 'cold' series is enough to change the behavioural responses to chemical signals.

Water Pollutants

Fish taste receptors are exposed to the environment and predisposed to the detrimental effects of water pollutants. It has been shown that many pollutants, especially heavy metals, affect fish taste reception by both destroying the taste buds and reducing the sensitivity to the taste stimuli. These events occur after a short exposure of the fish to polluted water.

Low pH Water

In natural water, with a pH between 7.6 and 7.8 fish demonstrated significant taste preferences to pellets containing all the classical taste substance used most of the 21 L- isomers of amino acids in the concentration range of 10^{-1} to 10^{-3} M were either ineffective or were deterrent stimuli.

References

Bardach, J. E., Fujiya, M. and Holl, A. 1965. Detergents: effects on the chemical senses of the fish *Ictalurus natalis* (leSeur). Science. 148: 1605-1607.

Carlos A. Ching. 2007. Chemoreception In Marine Shrimp,Nicovita-ALICORP SAA Technical Service. October – December pp: 1-3.

Carr, W. E. S. 1982. Chemical stimulation of feeding behavior. In: Chemoreception in Fishes (T. J. Hara, Ed.), Elsevier Scientific Publishing Co., Amsterdam. pp. 259-275.

Hara, T. J. 1976. Effects of pH on the olfactory responses to amino acids in rainbow trout, *Salmo gairdneri*. Comp. Biochem. Physiol. 54A: 37-39.

Heinen J.M. 1980. Chemoreception in decapod crustacea and chemical feeding stimulants as potential feed additives. *Proc. World Mar. Soc.* 11 : 319-344.

Fah, S.K. and Leng, Ch.Y. 1986. Some studies on the protein requirement of the guppy, *Peocilia reticulata* (Peters). J. Aquaricult. andAqua. Sci. 4 (4):325-335.

Liley, N. R. 1982. Chemical communication in fish. Can. J. Fish. Aquat. Sci. 39: 22-35.

Mackie, A. M. and Adron, J. W. 1978. Identification of inosine and inosine 5'-monophosphate as the gustatory feeding stimulants for the turbot, *Scophthalmus maximus. Comp. Biochem. Physiol*. 60A: 79-83.

Nunes, A.J.P. 2006. An update on the use of Chemo attractants in shrimp feeds. Aquaexpo 2006 – Guayaquil, Ecuador.

Part VII

Fish Breeding Technology

Chapter 13

Hormones in Reproduction of Ornamental Fishes

T. Francis

Department of Fisheries Biology and Capture Fisheries,
Fisheries College and Research Institute, Thoothukudi – 628 008, T.N.

The important functions of self-preservation (homeostasis) and species preservation (reproduction) in a multicellular organism demand the highest degree of integration of the various body systems and adaptation to external environment. The integration is brought about by the 'neuroendocrine complex' consisting of the 'Central Nervous System', the' Anterior Nervous System' and the 'Endocrine Glands'. These three work in close collaboration with each other and regulate the various vital functions such as circulation, respiration, digestion, excretion and reproduction.

Role of Endocrine Glands

The endocrine glands and the Anterior Nervous System serve to carry messages from the highest integrating authority, the CNS, to the body organs and cells. In turn, the Anterior Nervous System and the endocrine glands influence the functioning of the CNS. The endocrine glands consist of groups of highly specialized cells located at various sites in the body. They synthesize and discharge their specialized secretions directly into the circulation without the intervention of duct. These secretions are called Hormones. The hormones of the endocrine glands are peptides, steroids and catecholamines. Most of the hormones directly involved in fish reproduction are steroids and peptides. Steroids have structural similarities and are derived from cholesterol. Steroids are produced by the gonads and head kidney and are modified by the brain and liver. Peptide hormones are produced by the pituitary, brain and urophysis.

Testis

The androgens produced by the testis are the most important steroids for regulation of the male reproductive cycle. Androgens are C19 steroids and some fish androgens, such as testosterone, are also important androgens in higher vertebrates. But 11 Ketotestosterone, which may be unique to fish, is present in the plasma of some species in higher concentration than other androgens and is more active than testosterone in fish. Progestins (C21) and Estrogens (C18) are other steroids produced by the testis but their role in regulation of male reproductive function is not clear.

Ovary

The primary hormones secreted by the ovary are estrogens and progestins. The plasma of female channel catfish contains the following estrogens: 16 - Ketoestradiol, estriol and epiestroil. Plasma of female fish also contains testosterone (Lamba *et al.*, 1982). The ovarian estrogens production during vitellogenesis and progestins production during maturation seem to involve both granulosa and special thecal cells. For the synthesis of estrogens the thecal cells produce testosterone and then the granulosa cells metabolize the testosterone to 17 α estradiol.

Pituitary

Gonadotropin

Gonadotropin hormones secreted from pituitary are made up of glycoproteins (Gorman *et al.*, 1983). Some teleost fish have two types of gonadotropin hormone based on its carbohydrate content *i.e.*, Maturation hormone (high carbohydrate content and vitellogenic hormone (low carbohydrate content). The gonadotropin isolated from fish consists of two dissimilar sub-units that are similar in size to the gonadotropin sub-units of mammals. Each type of gonadotropin has a high molecular weight and a low molecular weight form a monomer-dimer relationship (Ng and Idler. 1978).

Common carp have only one type of each gonadotropin (Idler and Ng. 1979).Channel catfish has two types of gonadotropin (maturational gonadotrops and vitellogenic gonadotrops or gametogenic gonadotrops).

Thyrotropin

Thyrotropin is structurally similar to gonadotropin. In some species including channel catfish, thyrotrops can be distinguished from gonadotrops by their greater affinity for aldehyde fuchsion (Schreibman *et al.*, 1973; Massoud *et al.*, 1983). The region of pituitary where thyrotrops and gonadotrops are most abundant may also vary (Van Oordt and Peute, 1983).

Brain

There are at least two substances originating in the brain that regulate the release of gonadotropin. The first of these discovered in fish was gonadotropin releasing hormone (GnRH) which is similar to lutinizing hormone -releasing hormone (LHRH) of mammals (Sherwood *et al.*, 1984). The other neural chemical that influences the release of gonadotropin is Gonadotropin Release Inhibitory Factor (GRIF) (Peter, 1983). Dopamine may be an important gonadotropin release inhibitory factor (Chand *et al.*, 1983). The brain is also the site of aromatization of androgens to estrogen (Crim *et al.*, 1981).

Pineal

The pineal is connected to the dienocephalon of the brain and is a source of hormone Melatonin (Gorhman *et al.*, 1983). Melatonin is also synthesized in the retina of trout (Gem and Ralph, 1979).

Thyroid

The thyroid produces thyroxin and triodothyroxine (T3) in teleosts. One or both of these hormones are required for normal gonadal maturation.

Urophysis

The urophysis is part of the caudal neurosecretary system which includes specialized neurons in the posterior spinal cord (Grizzle and Rogers, 1976). Because of the direct neural connection between urophysis and the brain, the urophysis is probably controlled by brain. Several fractions that have hormonal properties have been isolated from the urophysis. These fractions are termed as urotensins. In carps, urotensin has four forms that differ in their amino-acid sequence (lchikwa *et al.*, 1984). In reproduction urotensin have the ability to cause contraction of smooth muscles in the sperm duct glandular testis and ovary.

Physiological Regulation of Reproduction

Production of Spermatozoa

Androgens stimulate spermatogenesis in fish (Figure 13.1) and also maintain glandular testis activity (Nayyar *et al.*, 1976; Fostier *et al.*, 1983). Androgens administered to fish may not have the same testicular effect as androgens produced in the testis because of the lack of transfer of plasma androgens from the blood to the somniferous tubules.

Spermiation may be triggered by an increase in androgen concentration. A rise in plasma androgens occurs at the time of spermiation. In some species (Schreck and Hopwood, 1974; Dindo and MacGregor. 1981; Campbell *et al.*, 1980; Schulz, 1984) and androgens injected into fish stimulate spermiation (Shehadeh *et al.*, 1973).

Development of Ovary

No hormones are required for *in vitro* proliferation of oogonia (Remacle *et al.*, 1976). The beginning of meiosis and previtellogenic growth of oocytes seems to be independent of pituitary hormones because meiosis occurs even after the removal of the pituitary (Tokarz, 1978). It is also found that gonadotropins also stimulate the formation of cytoplasmic vacuole. Involvement of both the gonadotropins and the process of yolk deposition have been demonstrated in some teleost species. Maturational hormone stimulates estrogen production by the ovary (Ng and Idler,

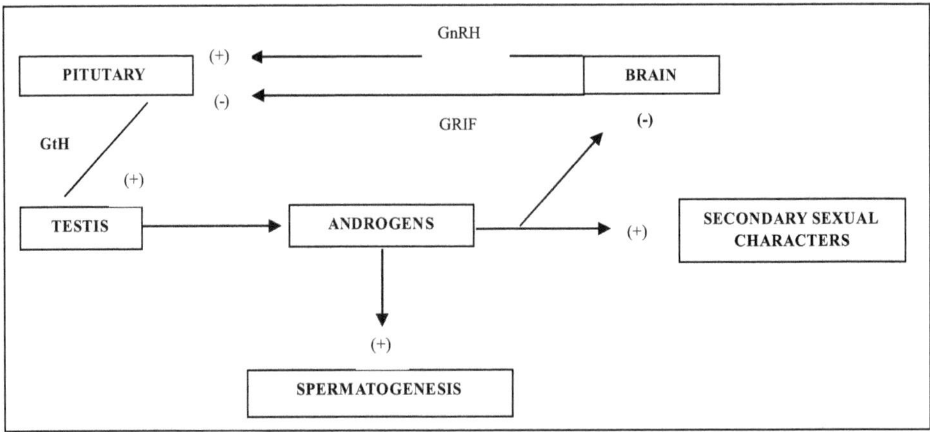

Figure 13.1: GtH – Gonadotropin; GnRH – Gonadotropin releasing hormone; GRIF – Gonadotropin release inhibitory factor.

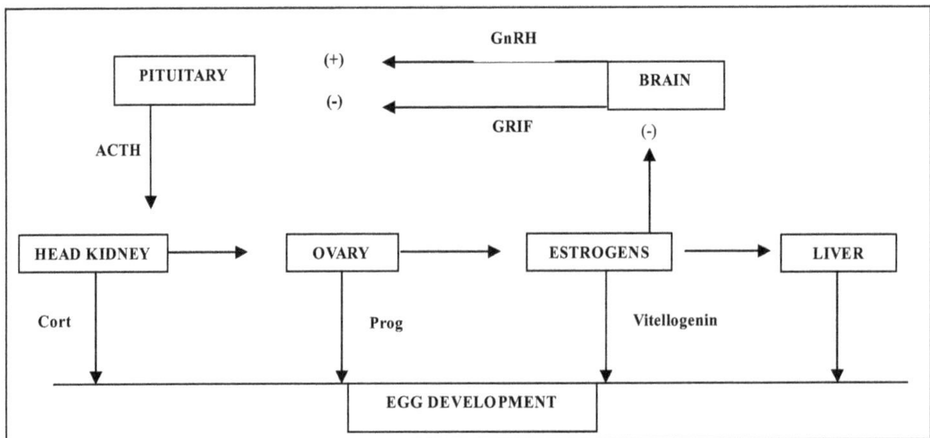

Figure 13.2: ACTH – Adrenocorticotropin; Cort – Corticosteroids; GEH – gonadotropin; GnRH – Gonadotropin releasing hormone; GRIF – Gonadotropin release- inhibitory factor; Prog – Progestron.

1983) and estrogens stimulate the liver to produce vitellogenin (Plack *et al.*, 1971; Hickey and Wallace).

The increased concentration of progestin and corticosteroids during the spawning season indicates that they have a role in oocyte maturation or ovulation. Prostaglandin concentrations in the ovary increases at the time of ovulation and prostaglandin stimulates *in vitro* follicle rupture of some species including gold fish (Stacey and Goetz, 1982). Indomethacin, an inhibitor of prostaglandin synthesis, blocks ovulation and injection of prostaglandin reverses the blockade.

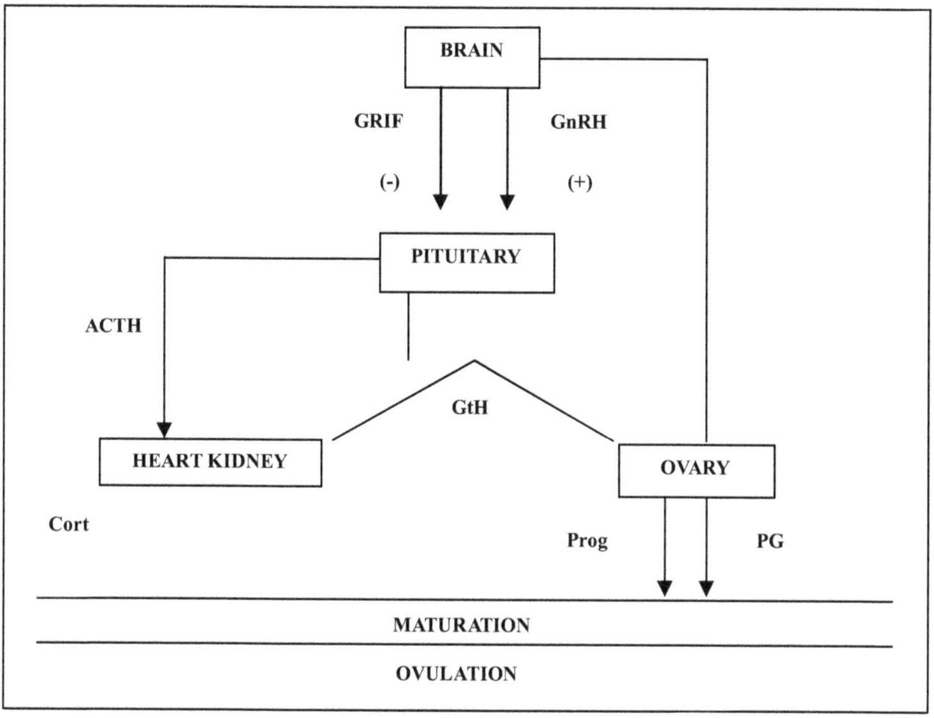

BRAIN

GRIF (-)

GnRH (+)

PITUITARY

ACTH

GtH

HEART KIDNEY

OVARY

Cort

Prog

PG

MATURATION

OVULATION

Figure 13.3: Ovulation Seems to be Controlled by Separate Mechanisms Involving Prostaglandins.

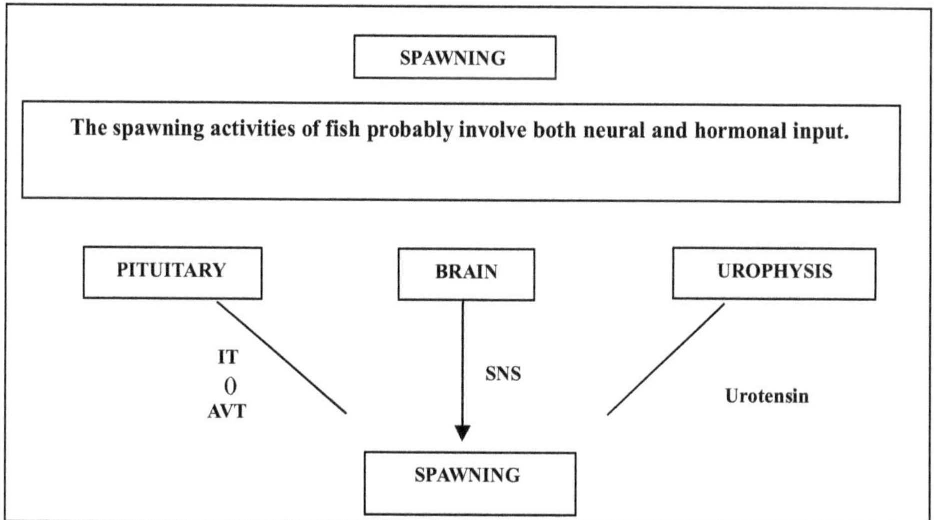

Figure 13.4: IT – Isotocin; AVT – Arginine vasotocin; SNS – Sympathetic nervous system.

Spawning

Role of Hormones in Achieving First Maturity

The role of steroid and gonadotropin hormones in achieving first maturity in fishes has been observed by Donaldson *et al.* (1972), Crim and Peter (1978) and Matre *et at.* (*1988*). It has been possible to achieve the first maturity (earlier than normally achieved in wild conditions) by hormonal treatments with both steroids and non-steroids easily in males than in females. Crim and Evans (1983) concluded that contrary to males, testosterone stimulation was not sufficient to stimulate onset of gonadal development in female fish.

Hormonal Applications for Advanced and Multiple Maturation and Breeding

A number of hormones such as human chorionic gonadotropin (HCG), synthetic luteinizing hormone releasing hormone (LHRHa) and mature pituitary gland solutions containing gonadotropin have been used to advance maturation and also achieve multiple maturation in Indian major carps (Singh, 1998, 2000). HCG injected intramuscularly @ 50 IU/kg/week for 4 weeks to prospective brood fishes of rohu advanced maturation to late April/early May. The action of HCG was mediated through pituitary gonadotropin accelerating the growth and maturation of oocytes (Das and Singh, 1990). Such fishes with advanced maturity and breeding in early May could be rematured at interval of 35-45 days while maintaining in good water quality and semi-balanced nutritious diet fortified with vitamins and minerals. The multiple (3-4 times) maturation and breeding of Indian major carps, rohu, catla and mrigal could be achieved nearly more than doubling the spawn production using same brood fish in one season (Anon, 1998, 1999). Advanced maturation of *Catla catla* was also achieved through improved diet and human chorionic gonadotropin injection (Somashekarappa *et al.*, 1990). It has been observed that the gonadotropin and steroid hormones in combination with other hormones such as thyroxine, prolactin, nutrition, water quality (specially O_2, CO_2, ammonia and pH), temperature and photoperiod play significant role in advancing maturity as well as attaining multiple maturation in carps and catfishes. The nutritious diet plays an important role in maturation of carps and catfishes (Somashekararppa *et. at.*, 1990; Singh, 1998, 2000; Pandey and Singh, 2003; Pandey *et al.*, 2003). The good water quality is one of the very important factor for maturation and rematuration 3-4 times in a season and induced breeding using ovaprim or hypophysation each time the fish attains full maturity (Singh, 2000).

It has been possible to advance maturation and attain multiple maturation and breeding in Indian catfish, *Heteropneustes fossilis* by human chorionic gonadotropin (HCG) injection or oral intubations and improved dietary feeding (Singh, 2000). The prospective brood fishes of singhi when injected with HCG @ 25 to 50 IU/kg or given oral intubations of HCG @ 25-50 IU/kg biweekly for 4-5 times advanced their maturation by more than 1-2 months (Kanungo *et al.*, 1999; Singh, 2000). The HCG treated matured fishes were bred in 3rd week of April or 1st week of May with fertilization rate ranging from 80-95 per cent and successful hatching of normal spawn. The same fish were rematured three times, in the same season, at an interval of 35-50 days and induced bred using ovaprim thus doubling the seed production (Singh, 2000).

The advanced and multiple maturation and breeding in catfish, *H. fossilis* by manipulating photoperiod-temperature regimes have been reported by Sundararaj and Vasal (1976). A photoperiod of 14L: 10D and water temperature between 25-30°C gave the best response for maturation and rematuration of ovary producing quality eggs and spawn on induced breeding (Sundararaj and Vasal, 1976).

Advancement of Maturation using Hormonal Pellets

Several studies have been made to implant hormonal pellets into the prospective brood fishes and advance their maturation and breeding ahead of normal breeding time. The hormones such as luteinizing hormone releasing hormone (LHRHa) are prepared in cholesterol pellets and implanted into the brood fish at the beginning of breeding season. The LHRH or HCG is slowly released from the pellet into the muscle from which it reaches the target organ stimulating growth of the gonad setting oogenesis and vitellogenesis in females and spermatogenesis in males much earlier and at faster rate (Almendras *et ai.*, 1988). The small amount of exogenous hormone from pellet is released regularly on sustained basis to the fish over a period of 1 to 1 ½ month. This accelerates the maturation process thus results in advancement of maturity. The detailed methodology for the preparation of hormonal pellets of LHRHa and HCG for implantation in carps and catfishes has been described by Singh (2000).

The implantation of hormonal pellet in prospective brood fishes of Indian major carps and catfishes is done at the beginning of preparatory phase of the breeding season (mid-February/early March) as the water temperature starts rising due to warming up. The brood fish is anaesthetized before implanting the pellet. It is important to select the site of pellet implantation which should be between the lateral line and the dorsal fin taking care not to damage the lateral line organs. In carps, a scale is removed and small incision (2-4) mm) is made into the muscle with sterilized blade, one pellet is then pushed into the muscle with probe which gets lodged into the incision being well covered with cut muscle preventing its contact with water as well as escape to the outside. If felt necessary the wound may be stitched with sutures preventing escape of the hormonal pellet. The wound heals after few days. The process of pellet implantation in catfish magur and singhi is similar eliminating the process of removal of scale since these have scale less skin (Singh, 2000).

The Indian major carps are implanted with LHRHa pellet @ 100 mg/fish in February -March to advance maturity by one month and these fishes were successfully bred by April-May producing normal larvae (Sahu, 1991). The advanced maturation was also observed in the Indian catfish, C. *batrachus* by implanting pellets of LHRHa @ 100 mg/fish during early part of March (Rao *et al., 1994*).

Role of GnRH and Dapamine-Antagonists in Fish Breeding

The presence of gonadotropin releasing hormone (GnRH) activity in the hypothalamus of teleosts is well documented (Breton *et. al.* 1971, 1972; Breton and Weil, 1973; Weil *et al.*, 1975). Prof. R.E. Peter and colleagues observed the presence of two hypothalamic hormones that control GtH release from pituitary (Peter, 1983).

Hypothalamus plays a great role in augmenting secretion of GtH in pituitary by secreting gonadotropin releasing hormone (GnRH) which is a decapetide (10 amino acids) quite similar to that known in mammals and other vertebrates. The molecular weight of GnRH is about 1,182 Da. Sherwood *et al.* (1983) have characterized fish gonadotropin releasing hormone (GnRH).

Hypothalamus also secretes another hormone, the dopamine, an amine which antagonizes the effect of GnRH in most of teleosts. It is now known that dopamine is or can mimic the gonadotropin release inhibitory factor (GnRIF). Studies conducted both *in vivo* and *in vitro* indicate that dopamine acts directly at the level of gonadotroph to inhibit GnRH stimulated GtH release in teleosts (Omiljaniuk *et al.*, 1987, 1989; Asselt *et al.*, 1988). It is also claimed that GnRH from Indian air-breathing teleost the murrel, *Channa punctatus* has been isolated purified and tested on the Indian major carps for successful induced breeding by Bhattacharya *et al.* (*1998*). They further claimed that GnRH from murrel in combination with domperidone gave better breeding response in the Indian major carps than ovaprim.

The efficiency of GnRH in releasing GtH from pituitary depends to a great extent on its combination with the chemical/compound blocking the action of GnRIF. The most commonly used dopamine antagonist, in combination with GnRH are (i) domperidone and (ii) pimozide. The other dopamine antagonists used are metaclopramide and centibutindol. Though metacloprarnide is soluble in water and commonly used in Israel in combination with GnRH, its potency is lower than domperidone and pimozide.

Domperidone was preferably used as dopamine antagonist by Lin *et al.* (1985, 1986, 1988) and Peter *et al.* (1988 a, b) in combination with sGnRHa and they developed an injection protocol of this combination for successful breeding of several species of fish. Their method of induced breeding was generally referred to as "Linpe" technique of induced ovulation and breeding for cultured freshwater fishes (Peter *et al.*, 1993; Pandey *et al.*, 1999).

References

Almenciras, I M., Duenas, C. Nacario, J. Sherwood, N. M. and Crin, L. W. 1988. Sustained hormone release. LU. Use of gonadotropin releasing hormone analogues to induce multiple s pawnings in sea bas, *Lates calcarifer. Aquaculture* 74: 97- 1 1 1.

Anon. 1998. "Fisheries management action plan for the Bahamas." *The Bahamas Reef Environment Educational Foundation* and *Macalister Elliott and Partners Ltd.*

Bhattacharya, S.K, Bhttacharya, A and Chakrabarti, A. 1998. Anxiogenic activity of intraventricularly administered arginine–vasopressin in the rat. *Biogenic Amines.* 14: 367–385.

Campbell, C. M., Fostier, A., Jalabert, B., and Truscott, B. 1980. Identification and quantification of steroids in the serum of rainbow trout during spermiation and oocyte maturation. *J. Endocrinol.* 85: 37 1 - 378.

Crim L. W. and Evans D. M. 1983. Influence of testosterone and or luteinizing hormone releasing hormone analogue on precocious sexual development in the juvenile rainbow trout. *Biol. Reprod.*, 29, 137-142.

Crim, L.W., Evans, D.M., Coy, D.H and Schally, A.V. 1981. Control of gonadotrophin hormones release in trout: Influence of synthetic LH-RH and LH-RH analogues *in vivo* and *in vitro. Life Sci.* 28: 129-135.

Dindo, J. J and MacGregor, R. 1981. Annual cycle of serum gonadal steroids and serum lipids in striped mullet. *Trans. Am. Fish. Soc.* 110: 403–409.

Fostier, A, Jalabert, R, Billard, R, Breton, Â and Zohar, Õ. 1983. The gonadal steroids; in *Fish physiology* (eds) W S Hoar, D J Randall and Å Ì Donaldson (New York: Academic Press) vol. 9, part A, 277-372.

Gern, W. A. and Ralph. C. L. 1979. Melatonin synthesis by the retina. *Science* 204:183–184.

Grizzle, J.M. and Roger, W.A. 1976. Anatomy and Histology of the channel catfish. Agric Exp. Station, Auburn Univ., Auburn, AL.

Ichikawa, Kiyoshi. 1984. Living environment and design of woonerf. *International Association of Traffic and Safety Sciences. 8: 40-51.*

Lamba, V.J., Goswami, S.V., and Sundararaj, B.J. 1982. Radioimmunoassay for plasma cortisol, testosterone, estradiol-17â and estrone in the catfish, *Heteropneustes fossilis* (Bloch): Development and validation. *Gen. Comp. Endocrinol.* 47: 170-181.

Lin, H. R., Van Der Kraak, G. Liang, J. Y., Peng, C., Li, G. Y., Lu, L. Z., Zhou, X. J., Chang, M. L. and Peter, R. E. 1986. The effects of dopamine on gonadotropin secretion and ovulation in fish cultured in China, p. 139-150. In R. Billard and J. Marcel (eds.) Aquaculture of cyprinids. INRA Service des publications, Paris.

Lin, H. R., Van Der Kraak, G., Zhou, X. J., Liang, J. Y., Peter, R. E., Rivier, J. E. and Vale, W. W. 1988. Effects of (D-Arg^6, Try^7, Leu^8, Pro^9 NEt)-leutinizing hormone-releasing hormone (LHRH-A), in combination with pimozide or domperidone on gonadotropin release and ovulation in the Chinese loach and common carp. *Gen. Comp. Endocrinol.* 69:31-40.

Nayyar, S.K., Keshavanath, P., Sundararaj, B.J and Donaldson, E.M. 1976. Maintenance of spermatogenesis and seminal vesicles in the hypophysectomized catfish, *Heteropneustes fossilis)*Bloch): Effects of ovine and salmon gonadotrophin and testosterone. *Can. J. Zool.*54: 285 – 292.

Ng, T. B. and Idler, D.R. 1978. Oe Biagn Doe Little forms of plaice vitellogenic and maturational hormones. *Gen. Comp. Endocrinol.* 34: 408 - 420

Pandey, A.C., Pandey, A.K. and Das, P. 1999. *Journal of Nature Conservation.* 11: 275-283.

Peter, R. E. 1983. The brain and neurohormones in teleost reproduction. In: Hoar WS, Randall DJ, Donaldson EM (eds), *Fish Physiology*, Vol 9, Academic Press, New York NY. 97–135.

Peter, R.E., Lin, H.R and Van der Kraak, G. 1988. Induced ovulation and spawning of cultures freshwater fish in ChINA: Advances in application of GnRH analogues and dopamine antagonist. *Aquaculture. 74: 1-10.*

Peter, R. E., Lin, H.-R., Van der Kraak, G. and Little, M. 1993. Releasing hormones, dopamine antagonists and induced spawning. p.25-30. *In:* JE. Muir and R.J. Roberts (eds.). Recent Advances in Aquaculture, vol. 4. B lackwell Science, Oxford.

Schreibman, M. P, Leatherland, J.F and McKeown, B.A. 1973. Functional morphology of the teleost pituitary gland. Am Zool. 13:719–742

Schulz, R., 1984. Serum levels of 11b-oxotestosterone in male and 17b-estradiol in female rainbow trout (Salmo gairdneri) during the first reproductive cycle. *Gen. Comp. Endocrinol.* 56, 111–120.

Sherwood, N., L. Eiden, M. Brownstein, J. Spiess, J. Rivier, and W. Vale.1983. Characterization of a teleost gonadotropin- releasing hormone. *Proc. Natl. Acad. Sci. U.S.A.* 80: 2794-2798.

Sherwood, Í. Ì., Harvey, Â, Browstein, Ì.J and Eiden, L.Å. 1984 Gonadotropin releasing hormone (GnRH) in stripped mullet (*Mugil cephalus),* milkfish (*Chanos chanos*) and Rainbow trout (*Salmo gairdneri*); comparison with salmon GnRH; *Gen. Comp. Endocrinol.* **55** 174-181.

Singh, B.N., Das, R.C., Sahu, A.K and Pandey, A.K. 2000. Balanced diet for the brood stock *Catla catla* and *Labeo rohita* and induced breedinh performance using ovaprim. *J. Advanced Zoology,* Central Institute of Freswater Aquaculture, India. 21(2): 92 – 97.

Singh, M. 1998. Regain of fertility and normality of progeny born during below protective threshold antibody titers in women immunized with the HSD hCG vaccine. *Am J Reprod Immunol.* 39:395–398.

Somashekarappa, B., Chandrashekaraiah, H.N and Nandeesha, M.C. 1990. Advancement of maturity in Indian major carp (*Catla catla*) through improved diet and human chorionic gonadotropin administration. Proc. of workshop on carp seed production technology. 2: 29-33.

Stacey, N. E and Goetz, F. W. 1982. Role of prostaglandins in fish reproduction. *Can. J. Fish. Aquat. Sci.* 39: 92-98.

Sundararaj, B.I. and Vasal, S. 1976 Photoperiod and temperature control in the regulation of reproduction in the female catfish, *Heteropneustes fossilis. J. Fish. Res. Board Can.,* 33:959-73.

Tokarz, R.R. 1978. Oogonial proliferation, Oogenesis and tolliculogenesis in non-mamammalian vertebrates. In. R.E. Jones (Ed), "The Vertebrate Ovary. Comparative Biology and Evolution". New York. Plenum Press. 145-179.

Van Oordt, P.G.W.J and Peute, J. 1983. The cellular origin of pituitary gonadolropins in teleosts. In: Hoar WS, Randall DJ, Donaldson EM (eds) *Fish Physiology* Vol IX Reproduction, Academic Press, New York, London, pp. 137–186.

Part VIII

Advanced Biotechnological Techniques in Aquariculture

Chapter 14

Cell Culture: Techniques and Applications

K. Kumanan

Department of Animal Biotechnology, Madras Veterinary College, Tamil Nadu Veterinary and Animal Sciences University, Chennai – 600 007

'Cell culture' is the *in vitro* propagation of cells harvested from animal/fish organs or tissues. There are three main methods of culturing cells *in vitro*. ORGAN CULTURE involves the maintenance of whole organ or parts of an organ in a way that may allow differentiation and preservation of the architecture and/or function. In PRIMARY EXPLANT/TISSUE CULTURE, fragments of tissues are placed at a glass-liquid interface where, following attachments, migration is promoted in the plane of the solid substrate. In CELL CULTURE, tissues or fragments of organs are dispersed either mechanically or enzymatically into a cell suspension and cultured as an adherent monolayer on a solid substrate or as a suspension in the culture medium. Out of these three basic methods, organ and primary explant culture methods are not widely used because of the difficulties experienced in maintaining these type of cultures for long periods. In the current scenario, culture of individual cells is the most preferred method. Under standard cell culture techniques, cells are immersed in a pool of medium containing essential nutrients and metabolic products. The concentration of both constituents changes as the cell population grows, thus influencing cell survival and function.

In- vitro culturing of cells could be discussed under two broad heading, namely primary cell cultures and continuous/established cell lines. In primary cell culture the tissues are disorganized by disrupting them into individual cells employing enzymes like trypsin and or collagenase. Such a disrupted tissue, apart from having specialized cells, will also contain connective tissue cells, blood cells and reticuloendothelial cells, most of which die in the subsequent cultures. However, if the cells multiply repeatedly for some more time, they can be subcultured and designated as secondary cultures. Some times the secondary cultures attain a potential to be subcultured indefinitely *in vitro* and such cells are designated as continuous cell lines.

Primary Cultures

Primary cell culture retain most of the characteristics of the cell from which they originated and generally they are anchorage dependant. These cells exihibit a phenomenon called 'CONTACT INHIBITION' which results in the cell, lining up in strongly oriented parallel strands. In any vessel these cells mutiply until they reach a maximum density after which they stop growing further. With regard to chromosomal number, these type of cells usually retain their diploid karyotype.

Continuous Cell Line

Continuous cell lines will have altered chromosomal number, shorter doubling time and attain immortalization. They do not exhibit the property of contact inhibition and some are anchorage independent and could be established in suspension cultures. Cell lines show a great variation in karyotype. Immediately after transformation, the incidence of tetraploid cells increase and subsequently anuploid cells make way with widespread chromosome numbers.

Animal Cell Culturing Techniques

One of the earliest experiments showing cell culture *in vitro* was conducted by Ross Harrison in 1907, who developed the 'hanging drop' method for culturing frog embryo tissues. In 1923, Carrel designed a flask suitable for routine cell culture work. The inadequacy experienced in the early days of animal cells in culture led the technologists to try imaginative methods of cell cultivation for getting better yields. As early as 1954, Earle and co-woarkers reported successful growth of the 'L' strain fibroblast using a conventional rotary shaker equipped with special flasks and media. Cherry and Hull (1956) are considered to be the fore runners of the present day suspension cultures. During the last two decades several new cultivation systems have been developed for mass production and concentration of animal cells. In general, these systems can be grouped into homogeneous bioreactor systems and heterogeneous bioreactor systems.

Homogeneous Bioreactor

In homogeneous bioreactor systems, the cells are distributed homogeneously in the reactor and this include all types of suspension cultures and microcarrier cultures. The classical suspension culture system for anchorage independent cells uses a stirred tank with different impeller types and installations equipped with or without a spin filter. In large scale bioreactors, modifications are generally made in the agitating systems. For culturing animal cells marine type impeller, vibromixer or rotating flexible sheets replace the turbine type impeller widely used in microbial fermentation. These agitation systems are aimed at reducing the strong shear forces that cause cell damage.

Novel bioreactors called air lift fermenters have the advantage of providing low turbulence and low shear force. Microcarrier (MC) culture is the growth or maintenance of anchorage dependant cells on small beads suspended in a stirred tank. MC cultures provide large surface area in relation to the volume of the vessel used.

Heterogeneous Bioreactors

These types of reactors are developed to accommodate the growth of surface adherent cells. Animal cells are cultured after adsorption to ceramic cylinders with uniform square channels along its length. This system is useful where production of a protein over a long period of time is necessary.

Entrapment of cells in various polymers like alginate, agarose, collagen and fibrin is a gentle means of immobilization which affords protection from mechanical stress. Encapsulation is another technology which involves two methods, one based on the membrane formation by cellulose sulfate (poly anion) and poly dimethyl diallyl ammonium chloride (poly cation). In all the capsules, the cells are separated from the medium by membranes.

Hollow fibre reactors consists of bundles of synthetic, semipermeable hollow fibres which offer a matrix for cell growth similar to the vascular system *in vivo*. Liquid can flow through the fibres (Intracapillary space) or through the space between the fibres (Extracapillary space). Generally, culture medium is pumped through the intracapillary space and a hydrostatic pressure permits the exchange of nutrients and waste products across the capillary wall. The cells and large molecular weight products are held in the extracapillary space. This sytem is suitable for both anchorage dependent and independent cells.

Likewise, packed bed reactors consists of a static bed of solid inert particles such as glass beads with a typical diameter of 3-5 mm and cells are allowed to attach and grow on the bead surface. The medium is re circulated through the packed bed by a pump and oxygenated by an input of air in a secondary vessel.

Applications of Animal Cell Culture

Although cell culture systems have been used for growing viruses, it was not until the late 1940s that this system became widely used among virologists. Its wide spread application was initiated after the observation of Enders and his co-workers in 1949 that poliomyelitis virus would grow in human tissue of non-nervous origin in cell culture. Cell cultures have a wide range of applications with several biological products used in the diagnosis and control of infectious diseases being produced through cell culture.

1. Disease Diagnosis

Isolation and subsequent identification of the causative agent is the only way to prove the etiology of any viral infection. Since viruses require a living system for their replication, the virologists are left with only three options, employing either experimental animals, embryonated eggs or cell culture systems. Both primary cultures and cell lines are routinely used for virus isolation and identification. Cell culture systems also offer a wonderful platform for pathogenicity studies of fish and animal viruses.

2. Viral Vaccines

Apart from the diagnostic aspects, cell culture system plays a vital role in the development of viral vaccines. These viral vaccines seem to be the main animal cell product with high market value. Animal cells are used as the substrate for vaccine production. *In vitro* cultured animal or fish cells can be used for attenuating virulent viruses for developing live or killed vaccines.

3. Monoclonal Antibodies

Somatic cell hybrids have become an important source of cellular products which cannot be obtained from short primary cultures. The best example for such system is the hybrid myeloma used in the production of monoclonal antibodies against the antigen of choice. The production of monoclonal antibodies involves the inoculation of the antigen of choice into BALB/c mice, the spleen cells of which could be removed from the mice and put into culture. These cells in culture will divide and produce antibodies directed against the antigen but will last only for a short period in culture. However, cell hybridization makes it possible to combine the spleen cells producing antibodies with mouse myeloma cells which can be cultured continuously. The mouse myeloma cells namely SP_2/o cells are fused with the spleen cells and the resulting hybrids are screened for the production of desired antibodies. Such hybrids will provide a continuous supply of monoclonal antibodies which can be used as reagents in diagnostic techniques.

4. Hormones

Glycoprotein hormones are produced in *in vitro* cultured cells and have therapeutic use. Several clones of pituitary tumor cells are available which synthesize adrenocorticotropic hormone. These include the rat GH cells and mouse AtT20/D16 cells. The GH cells previously known as MtTW5 were established from rat pituitary tumor. These cells apart from producing growth hormone, also synthesize prolactin, the ratio of which varies according to the passage level. The AtT20/D16 cells are clonal strains isolated from a radiation induced pituitary tumor in mouse. Islets of langerhans of rat, pig and human foetus are grown *in vitro* for the production of insulin employing hollow fibre cell culture system.

5. Immunoregulators

Animal cells are employed in the production of several non-antibody immunoregulators like interferons, interleukins, colony stimulating factors, B-cell growth factor, macrophage activating factor, T-cell replacing factor and migration inhibition factor.

6. Tumor Specific Antigens

The best example of this is the carcino-embryonic antigen (CEA). CEA is a tumor associated antigen produced by adenocarcinoma cells which when found in sera or other body fluids indicates a state of malignancy. Production of CEA is achieved by its extraction from cultured adenocarcinoma cell lines. The tumor specific antigen is very useful in screening for antitumor agents and detecting malignancy in patients using immuno assays.

7. Reconstitution of Living Skin

Attempts are being made to employ animal cell technology for reconstitution of living skin. Reconstituted skin comprises of two components namely, a dermal equivalent made up of fibroblasts in a collagen matrix and an epidermal equivalent developed from karatinocytes. The dermal equivalent tissues are accepted as allografts.

8. *In-vitro* Toxicity Assays

Cell lines like BHK_{21} HeLa, 3T3 and L929 are used in *in-vitro* toxicity assays. Long term cytotoxicity assay involves exposure of 3T3-L1 cells to relatively low concentrations of test materials (in ug) for upto 72 hrs with measurement of final total cellular proteins by kenacid blue staining as a means of evaluating cell proliferation inhibition. Neutral red release method is employed in short term cytotoxicity assays.

9. Expression of Recombinant Proteins

Cell lines like SF9, CHO, CV-1 and BS-C-1 are used in the expression studies of recombinant proteins.

10. *In-vitro* Culturing of Embryos

In vitro cultured oviductal epithelial cells and cumulus cells are used as feeder layers for *in vitro* maturation of oocytes and culturing of embryos.

11. Fish Cell Lines

Introduction

Tissue culture and the development of cell lines from fish are priorities for pathogen detection, toxicological studies, carcinogenesis, cellular physiology and genetic regulation and expression studies. The first fish cell line, designated as RTG-2, was developed in 1962 from gonad of rainbow trout (*Oncorhynchus mykiss*) and even today this cell line has tremendous applications in virological and toxicological studies (Wolf and Quimby, 1962). Since then the work on developing fish cell lines has progressed and the number of fish cell lines increased tremendously to 283 (Lakra *et al.*, 2011a). As of today our scientific community does not have a complete list of cell lines established from Indian fishes. And therefore, this compilation provides complete information on fish cell lines established through DBT funded R&D projects carried out in different laboratories in India. The purpose of compiling a list of cell lines of Indian fishes is to keep the scientific community abreast of information regarding the current availability of cell lines. This publication highlights the potential applications of the rich resource of cell lines available for research and teaching. It will also help to avoid the development of cell lines which are already available.

National Status on Fish Cell Lines

The information on the cell lines/culture from fish is quite extensive in the advanced countries but the information in this field from India is limited as evidenced in the literature especially in the application of fish cell lines. Most of the works have been mainly based on primary cell cultures till the publication of the work by Sahul Hameed *et al.* (2006) on permanent fish cell lines. The work on fish cell culture was initiated in India under the FAO/UNDP project entitled "Intensification of Freshwater Aquaculture" during the early eighties. A modest fish cell culture facility was established at CIFA (the then FARTC of CIFRI), Bhubaneswar under its newly created Division of Ichthyopathology and Fish Health Protection. The institute received overseas training at Veterinary Faculty, University of Zagreb in the maintenance and sub - cultivation of EPC, RTG - 2 and FHM cell lines, experimental infection of EPC cell lines by SVCV, observation of CPE, large scale production of vaccine against SVC and participation in large scale vaccination of yearlings of common carps in commercial farms. Initial success was in the form of obtaining primary monolayer culture of fish cells using both explant of minced tissues and seeding of trypsin mediated dispersed cells from tissues such as kidney, heart, fin and tumor cells. In the mid Eighties, four cell lines namely BB, FHM, RTG-2 and EPC were maintained. Attempts were also made to develop cell lines from carps and a cell line was developed from silver carp tumor cells which underwent over fifty passages (Anonymous, 1989, 1990, 1991). Unfortunately there was catastrophic loss of all cultures due to serious contamination and the work was suspended for nearly four to five years. Subsequently fish cell culture work got a boost in India with the support of DBT.

Singh *et al.* (1995) developed primary cell culture from kidney of freshwater fish (*Heteropneustus fossilis*). Sathe *et al.* (1995) established and characterized a new fish cell line (with 30 passages only), MG-3 from the gills of mrigal, (*Cirrhinus mrigala*). A cell line was developed from the gill tissues of Indian cyprinoid (*Labeo rohita*) (Sathe *et al.*, 1997). Lakra and Bhonde (1996) developed primary cultures from the caudal fin of an Indian major carp, (*Labeo rohita*). Rao *et al.* (1997) developed primary cell culture from the heart of Indian major carps.

Prasanna *et al.* (2000) initiated cell culture from fin explant of mahseer (*Tor putitora*). Sunil Kumar *et al.* (2001) developed primary cell culture system from the ovarian tissue of African catfish which passaged only 15 times after which they ceased to multiply and consequently perished. Lakra *et al.* (2005) developed two new primary cell culture systems from fry and fingerling of the (*Lates calcarifer*). Lakra *et al.* (2006) developed a new fibroblast like cell line from the fry of golden mahseer *Tor putitora* (Ham). A continuous cell line (SISK) from kidney of sea bass, *L. calcarifer*, has been established, characterized and designated as SISK (Sahul Hameed *et al.*, 2006). The SISK cell line is India's first marine fish cell line. The development and characterization of a new tropical marine fish cell line (SISS), derived from the spleen of sea bass, *L. calcarifer* has been described (Parameswaran *et al.*, 2006a). A continuous embryonic stem (ES) cell line was established from blastula stage embryos of sea bass (*L. calcarifer*) (Parameswaran *et al.*, 2006b). The ES cells were round or polygonal and grew exponentially in culture. Two new cell lines, SIMH and SIGE, were most successfully developed from the heart of milkfish (*Chanos chanos*), a euryhaline teleost, and from the eye of grouper (*Epinephelus coioides*), respectively (Parameswaran *et al.*, 2007a). The SBES cells exhibited an intense alkaline phosphatase activity. On treatment with all-trans retinoic acid, the cells were differentiated into neuron-like cells, muscle cells and beating cardiomyocytes indicating their pluripotency (Parameswaran *et al.*, 2007b). A new cell line [Sahul India Catla Eye (SICE)] has been developed from eye tissue of Indian major carp (*Catla catla*), a freshwater fish cultivated in India (Ishaq Ahmed *et al.*, 2008). The SICE cell line consists predominantly of epithelial-like cells. These cells are strongly positive for epithelial markers such as pancytokeratin and cytokeratin 19. Two new cell lines, designated RE and CB, were developed from the eye of rohu, *L. rohita*, and the brain of catla, *C. catla*, respectively (Ishaq Ahmed *et al.*, 2009a). The RE cell line was sub-cultured for more than 70 passages and the CB cell line for more than 35 passages.

The RE cells are rounded and consist predominantly of epithelial cells while the CB cell line consists of predominantly fibroblastic-like cells. Ishaq Ahmed *et al.* (2009b) developed a new cell line (SICH) from heart muscle of Indian major carp (*C. catla*) and this cell line has been sub-cultured more than 130 times in a period of two years. The SICH cell line consists predominantly of fibroblastic-like cells. Babu *et al.* (2011) developed a new cell line, Indian Catfish Fin, derived from the fin tissue of Indian walking catfish, *Clarias batrachus*, which was established and characterized. The cell line has been subcultured more than 110 times since its initiation in 2007.

A primary cell culture has been developed from heart tissue of *E. tauvina* and gill tissue of *E. malabaricus* (Sobhana *et al.*, 2008, 2009). Swaminathan *et al.* (2010) developed and characterized a new cell line from caudal fin of *Etroplus suratensis*. Lakra *et al.* (2010a) developed three new cell lines from fin, heart, and swim bladder of *L. rohita* and evaluated them for optimal growth conditions and also carried out molecular and cytogenetic authentication of these cell lines. Two new cell lines designated as CCF and CCH were established, respectively from fin and heart tissues of common carp, *Cyprinus carpio* (Lakra *et al.*, 2010b). Similarly, two cell lines namely PDF and PDH were developed from the caudal fin and heart of *P. denisonii*, respectively (Lakra *et al.*, 2011b). Lakra and Goswami (2011) developed and characterized a continuous cell line from the caudal fin of *Puntius sophore*. Swaminathan *et al.* (2012) developed a fibro-blastic-like cell line from the ornamental fish, *Puntius denisonii*. Cell culture systems were also developed from different tissues such as eye, fin, heart and swim bladder of *P. chelynoides* using explant method (Goswami *et al.*, 2012). Four novel cell lines from tissues of eye, gill, kidney and brain of *E. suratensis* were developed and characterized (Babu *et al.*, 2012). The cell lines of eye, gill, kidney and brain had been sub-cultured for 245, 185, 170 and 90 passages, respectively, since 2008. These cell lines exhibited predominantly epithelial-like cells.

Uma *et al.* (2002) developed a primary cell culture system from the hepatopancreas of *Penaeus monodon* and maintained it up to a maximum period of 12 weeks by subculturing seven times. Sashikumar and Desai (2008) reported the development of primary cell culture from hepatopancreas of edible crab, *Scylla serrata* using crab saline, L-15 (Leibovitz), 1 X, L-15 + crab saline, 2 X, L-15 + crab saline, 3 X L-15 and citrate buffer without any serum. Likewise, Goswami *et al.* (2010) developed a primary culture from heart tissues of *Macrobrachium rosenbergii* by explant culture technique.

A primary hemocyte culture from *P. monodon* had been developed for assessing the titre value of white spot syndrome virus and cytotoxicity and genotoxicity of heavy metals (Jose *et al.*, 2010, 2011). Primary cell culture from lymphoid tissues of *P. monodon* was developed to demonstrate the expression of WSSV and WSSV-induced immune-related genes (Jose *et al.*, 2012). The heart explants derived from freshwater crab (*Paratelphusa hydrodomous*) were exposed to WSSV and the infection was confirmed by PCR, RT-PCR, Western blot, histology, immunohistochemistry, bioassay and transmission electron microscopy (Nathiga Nambi *et al.*, 2012). The WSSV was quantified by real-time PCR and indirect ELISA, and the results indicated that the heart explants of *P. hydrodomous* would be a good *in vitro* system for WSSV replication.

Application of Fish Cell Lines

Fish cell lines have both fundamental and practical applications. A diversity of questions can be addressed with fish cell cultures. These include applications in biomedical research, toxicology and basic fish research, viral isolation, gene regulation, expression and gene transfer. The most widely employed application of fish cell lines is the isolation of fish viruses that are agents of epizootics of economically important cultured fish species. New cell lines from a variety of fish species are being developed and used for the isolation and characterization of fish viruses including previously unknown viruses. In addition to serving solely as a means of virus isolation, fish cell cultures are very useful as *in vitro* models for studying the replication and genetics of these viruses, the establishment and maintenance of virus carrier states, the effect of antiviral drugs, and in the production of experimental vaccines. Indian fish cell lines have been tested for their susceptibility to different fish viruses such as nodavirus, BSNV, LCNNV and IPNV. Fish nodavirus responsible for severe mortality in seabass hatchery was isolated for the first time in India using sea bass kidney cell line (Parameswaran *et al.*, 2008) and whole nodavirus vaccine had been produced on large-scale using spleen cell line of sea bass. Susceptibility of the lymphoid cells derived from *P. monodon* to WSSV was tested and the infection was confirmed by immunofluoresence assay using monoclonal antibody raised against the 28 kDa envelope protein of WSSV (Jose *et al.*, 2012). The heart explants of freshwater crab can be used as *in vitro* system for production of WSSV and can also be used for evaluating anti-WSSV substances against WSSV (Nathiga Nambi *et al.*, 2012). Suthindhiran *et al.* (2011) demonstrated the antiviral activity of furan-2-yl acetate ($C_6H_6O_3$) extracted from *Streptomyces* VITSDK1 spp. in eye cell line of grouper infected with fish nodavirus.

Primary cell cultures form an important tool in understanding the fundamental aspects of fish immunology (Faisal and Ahne, 1990). Primary cultures have been used for studies on the mitogenic response of fish lymphocytes (Faulman *et al.*, 1983), the *in vitro* generation and detection of antibody-secreting cells (Miller and Clem, 1984; Kaattari *et al.*, 1986). Macrophage cultures have been used to study phagocytosis (Sovenyi and Kusuda, 1987), the respiratory burst (Chung and Secombes, 1988), lipoxin synthesis (Pettitt *et al.*, 1989), t he formation of multinucleate giant cells (Secombes, 1985), bactericidal activity (Olivier *et al.*, 1986), and association with fish pathogens (Trust *et al.*, 1983). Such studies need to be carried out in Indian fish using respective cell lines established from different fish species of India.

Fish cell cultures are finding application in toxicology for evaluating effects of various chemicals, pesticides and industrial wastes and in the study of carcinogenesis as *in vitro* models for investigating cell transformation by fish viruses, chemical agents, and the interaction of viruses and chemical carcinogens. Cell lines of *E. suratensis* established were evaluated for their potential use as screening tools for the ecotoxicological assessment of tannery effluent and the results of *in vitro* studies were correlated with the results of *in vivo* lethality test using fish. The use of these cell lines for toxicity assessment of industrial effluents instead of living fish was recommended (Taju *et al.*, 2012). More studies need to be carried out to make use of established Indian fish cell lines in toxicological studies.

Table 14.1: Indian Investigators Working on Fish Cell Lines

Sl.No.	*Name and Address of Investigator*	*No. of Fish Cell Lines Available*
1.	**Prof. I.S. Bright Singh** National Centre for Aquatic Animal Health, Cochin University of Science and Technology, Lakeside campus, Fine Arts Avenue, Cochin -682016. Phone No. : 0484-2381120; Fax +91- 484 2381120 E-mail : isbsingh@gmail.com; http://www.ncaah.org	3
2.	**Dr. M. Goswami** Senior Scientist, National Bureau of Fish Genetic Resources, Canal Ring Road, PO Dilkusha, Lucknow- 226002, India Fax: +91 522 2442441 E-mail: mukugoswam@gmail.com	3
3.	**Dr. W.S. Lakra** Director, Central Institute of Fisheries Education Versova, Andheri (W), Mumbai - 400 061 Phone No.: +91-22-26363404; +91-22- 26361446 E-mail: wslakra@cife.edu.in http://www.cife.edu.in	8
4.	**Dr. K. Riji John** Professor, Department of Aquaculture, Tamilnadu Fisheries University, Fisheries College and Research Institute, Thoothukkudi, Tamilnadu, India 628008 Phone No.: 061-2340554 (O) E-mail: rijijohn@gmail.com	3
5.	**Dr. A. Sait Sahul Hameed** OIE Expert and Associate Professor, Department of Zoology, C. Abdul Hakeem College, Melvisharam – 632 509, Vellore Dt., Tamil Nadu Ph. Office : 04172 - 269487; Res : 04172 - 233730 E-Mail: cah_sahul@hotmail.com	24
6.	**Dr. Sobhana Pradeep** Senior Scientist, Marine Biotechnology Division, Central Marine Fisheries Research Institute, Kochi - 682 018, India E-mail: sobhana_pradeep@yahoo.co.in	13

Table 14.2: International Repositories for Fish Cell Lines

Sl.No.	Name and Address of Repository	No. of Fish Cell Lines Available
1.	**American Type Culture Collection (ATCC)** P.O. Box 1549, Manassas, VA 20108, USA. Tel: (703) 365-2700; Fax: (703) 365-2701 E-mail: help@atcc.org; URL: http://www.atcc.org/	19
2.	**Cell Bank Australia, Australia** E-mail: info@cellbankaustralia.com; www.cellbankaustralia.com	16
3.	**European Collection of Cell Cultures (ECACC)** CAMR Centre for Applied Microbiology and Research Porton Down, Salisbury, Wiltshire (UK),SP4 0JG, UK. Tel.: +44 (0) 1980 612512; Fax: +44 1980 611315 E-mail: ecacc@camr.org.uk; URL: http://www.ecacc.org.uk/	22
4.	**German Collection of Microorganisms and Cell Cultures (DSMZ)** Mascheroder Weg 1b, 38124 Braunschweig, Germany. Tel.: þ49 (0) 531-2616-0; Fax: þ49 (0) 531-2616-418 E-mail: help@dsmz.de; URL: http://www.dsmz.de/index.html	4
5.	**Istituto Zooprofilattico Sperimentale (IZSBS)** Centro Substrati Cellulari Via A.Bianchi 7, Brescia (BS) 25100, Italy Tel.:+39-0302290248; Fax.: +39-030225613 E-mail:substr@bs.izs.it, URL: http://www.biotech.ist.unige.it/cldb/descat6d.html	12
6.	**Riken Cell Bank (RIKEN)** The Institute of Physical and Chemical Research, 3-1-1 Koyadai, Tsukuba Science City, Ibaraki 305, Japan, Tel.: +81 -298-36-3611, Fax.: +81 -298-36-9130 E-mail:cellbank@riken.go.jp, URL: http://www.rtc.riken.go.jp/.index.html	12

Table 14.3: Details of Indian Fish Cell Lines Developed

Sl.No.	Fish Species	Organ	Designation	Cell Type	Maximum Passage Obtained	Application
1.	*Lates calcarifer*	Kidney	SISK	Epithelial	135	Virological
2.	*Lates calcarifer*	Spleen	SISS	Fibroblastic	197	Virological
3.	*Epinephelus coioides*	Eye	SIGE	Epithelial	204	Virological
4.	*Epinephelus coioides*	Kidney	GK	Fibroblastic	130	Nodavirus
5.	*Epinephelus coioides*	Heart	GH	Fibroblastic	78	–
6.	*Epinephelus coioides*	Brain	GB	Neuronal	60	Neuro Toxicology
7.	*Catla catla*	Heart	SICH	Fibroblastic	215	Toxicology
8.	*Catla catla*	Brain	CB	Fibroblastic	92	Neuro Toxicology
9.	*Catla catla*	Eye	SICE	Fibroblastic	115	Toxicology
10.	*Catla catla*	Gill	ICG	Fibroblastic	48	Toxicology

Contd...

Table 14.3–*Contd...*

Sl.No.	Fish Species	Organ	Designation	Cell Type	Maximum Passage Obtained	Application
11.	*Labeo rohita*	Eye	RE	Epithelial	90	–
12.	*Labeo rohita*	Gill	RG	Epithelial	62	Toxicology
13.	*Clarius batrachus*	Fin	ICF	Fibroblastic	110	Virological
14.	*Etroplus suratensis*	Eye	IEE	Epithelial	245	Virological
15.	*Etroplus suratensis*	Gill	IEG	Epithelial	120	Virological
16.	*Etroplus suratensis*	Kidney	IEK	Epithelial	185	Virological
17.	*Etroplus suratensis*	Brain	IEB	Fibroblastic	90	Neuro Toxicology
18.	*Danio rerio*	Eye	DRE	Epithelial	82	–
19.	*Danio rerio*	Fin	DRF	Fibroblastic	78	–
20.	*Danio rerio*	Gill	DRG	Epithelial	48	–
21.	*Channa striatus*	Kidney	CSK	Fibroblastic	85	–
22.	*Channa striatus*	Gill	CSG	Epithelial	75	–
23.	*Channa striatus*	Heart	CSH	Epithelial	48	–
24.	*Cyprinus carpio carpio*	Eye	KCE	Epithelial	44	–
25.	*Rachycentron canadum*	Brain	RC3BrTr	Epithelial	74	–
26.	*Rachycentron canadum*	Caudal Peduncle	–	Epithelial	72	–
27.	*Rachycentron canadum*	Fin	RC4F2Tr	Fibroblastic	100	–
28.	*Rachycentron canadum*	Heart	RC4H1Tr	Fibroblastic	99	–
29.	*Dascyllus trimaculatus*	Caudal Peduncle	DT1CpEx	Fibroblastic	138	–
30.	*Dascyllus trimaculatus*	Fin	DT1F4Ex	Fibroblastic	146	–
31.	*Epinephelus merra*	Brain	HC2BrTr	Fibroblastic	54	–
32.	*Epinephelus merra*	Caudal Peduncle	HC2CpTr	Epithelial	77	–
33.	*Epinephelus merra*	Gill	HC2G1Tr	Fibroblastic	85	–
34.	*Epinephelus merra*	Heart	HC2H2Ex	Epithelial and Fibroblastic	86	–
35.	*Epinephelus merra*	Spleen	HC2SpEx	Epithelial	70	–
36.	*Epinephelus merra*	Fin	HC2F3Ex	Epithelial	63	–
37.	*Siganus canaliculatus*	Brain	SC12Br1Tr	Epithelial	34	–
38.	*Amphiprion sebae*	Fin	CFFN	Fibroblastic	178	Virological
39.	*Amphiprion sebae*	Brain	CFBR	Fibroblastic	121	–
40.	*Amphiprion sebae*	Caudal Peduncle	CFCP1	Fibroblastic	133	Virological
41.	*Cyprinus carpio*	Fin	CCF	Epithelial	76	–

Contd...

Table 14.3–*Contd...*

Sl.No.	*Fish Species*	*Organ*	*Designation*	*Cell Type*	*Maximum Passage Obtained*	*Application*
42.	*Cyprinus carpio*	Heart	CCH	Fibroblastic	61	–
43.	*Puntius chelynoides*	Eye	PCE	Fibroblastic	31	–
44.	*Puntius denisonii*	Fin	PDF	Fibroblastic	74	–
45.	*Puntius sophore*	Fin	PSCF	Fibroblastic	116	–
46.	*Labeo rohita*	Fin	RF	Fibroblastic	63	–
47.	*Labeo rohita*	Heart	RH	Fibroblastic	58	–
48.	*Labeo rohita*	Swim Bladder	RSB	Epithelial	56	–
49.	*Schizothorax richardsonii*	Fin	SRCF	Fibroblast	55	–
50.	*Schizothorax richardsonii*	Eye	SRE	–	11	–
51.	*Tor tor*	Fin	TTCF	Fibroblastic	74	–
52.	*Penaeus monodon*	Lymphoid	–	Epitheloid	Primary cell culture	Virological
53.	*Penaeus monodon*	Lymphoid	–	Fibroblastic	Primary cell culture	Virological
54.	*Penaeus monodon*	Haemocytes	–	Fibroblastic	Primary cell culture	Virological
55.	*Paratelphusa hydrodomous*	Heart	–	–	Explant culture	Virological

References

Anonymous. 1989. Annual Report of Central Institute of Freshwater Aquaculture, Bhubaneswar, Orissa.

Anonymous. 1990. Annual Report of Central Institute of Freshwater Aquaculture, Bhubaneswar, Orissa.

Anonymous. 1991. Annual Report of Central Institute of Freshwater Aquaculture, Bhubaneswar, Orissa.

Babu, V.S., Chandra, V., Nambi, K.S N., Majeed, S. A., Taju, G., Patole, M.S. and Sahul Hameed, A.S. 2012. Development and characterization of novel cell lines from *Etroplus suratensis* and their applications in virology, toxicology and gene expression. *Journal of Fish Biology*. 80: 312-334.

Babu, V.S., Nambi, K.S.N., Chandra, V., Ahmed, V.P.I., Bhonde, R. and Sahul Hameed, A.S. 2011. Establishment and characterization of a fin cell line from Indian walking catfish, *Clarias batrachus* (L.). *Journal of Fish Diseases*. 34: 355-364.

Chung, S. and Secombes, C.J. 1988. Analysis of events occurring within teleost macrophages during the respiratory burst. *Comparative Biochemistry and Physiology*. 89B: 539-544.

Faisal, M. and Ahne, W. 1990. A cell line (CLC) of adherent peripheral blood mononuclear leucocytes of normal common carp, *Cyprinus carpio*. *Developmental and Comparative Immunology*. 14: 255-260.

Faulmann, E., Cuchens, M.A., Lobb, C.J., Miller, N.W. and Clem, L.W. 1983. An effective culture system for studying *in vitro* mitogenic responses of channel catfish lymphocytes. *Transantlatic American Fisheries Society*. 112: 673-679.

Goswami, M., Lakra, W.S., Rajaswaminathan, T. and Rathore, G. 2010. Development of cell culture system from the giant freshwater prawn *Macrobrachium rosenbergii* (de Man). *Molecular Biology Reports*. 37: 2043-2048.

Goswami, M., Sharma, B.S., Tripathi, A.K., Kamalendra Yadav, Bahuguna, S.N., Nagpure, N.S., Lakra, W.S. and Jena, J.K. 2012. Development and characterization of cell culture systems from *Puntius* (Tor) *chelynoides* (McClelland). *Gene*. 500: 140-147.

Ishaq Ahmed, V.P., Chandra, V., Sudhakaran, R., Rajesh Kumar, S., Sarathi, M., Sarath Babu, V., Ramesh, B. and Sahul Hameed, A.S. 2009a. Development and characterization of cell lines derived from rohu, *Labeo rohita* (Hamilton), and catla, *Catla catla* (Hamilton). *Journal of Fish Diseases*. 32: 211-218.

Ishaq Ahmed, V.P., Sarath Babu. V., Vikash Chandra, Nambi, K.S.N., John Thomas, Ramesh Bhonde. and Sahul Hameed. A.S. 2009b. A new fibroblastic - like cell line from heart muscle of the Indian major carp (*Catla catla*): Development and characterization. *Aquaculture*. 293:180-186.

Ishaq Ahmed. V.P., Vikash Chandra, Parameswaran. P., Venkatesan, C., Ravi Shukla, Ramesh Bhonde. and Sahul Hameed. A.S. 2008. A new epithelial-like cell line from eye muscle of catla *Catla catla* (Hamilton): development and characterization. *Journal of Fish Biology*. 72: 2026-2038.

Jose, S., Jayesh, P., Mohandas, A., Rosamma Philip. and Bright Singh, I.S. 2011. Application of primary haemocyte culture of *Penaeus monodon* in the assessment of cytotoxicity and genotoxicity of heavy metals and pesticides. *Marine Environmental Research* doi:10.1016/j.marenvres.2010.12.008.

Jose, S., Jayesh, P., Sudheer, N.S., Poulose, G., Mohandas, A., Philip, R. and Bright Singh, I.S. 2012. Lymphoid organ cell culture system from *Penaeus monodon* (Fabricius) as a platform for white spot syndrome virus and shrimp immune-related gene expression. *Journal of Fish Diseases* doi:10.1111/j.1365-2761.2012.01348.x.

Jose, S., Mohandas, A., Philip, R. and Singh, I. 2010. Primary haemocyte culture of *Penaeus monodon* as an *in vitro* model for white spot syndrome virus titration, viral and immune related gene expression and cytotoxicity assays. *Journal of Invertebrate Pathology*, 105: 312-321.

Kaattari, S., Blaaustein, K., Turaga, P., Irwin, M. and Weins, G. 1986. Development of a vaccine for bacterial kidney disease in salmon. Annual Report for 1985 U. S. Dept. Energy Bonneville Power Admin. Division of Fish &Wildlife portal.

Lakra, W.S., Behera, M.R., Sivakumar, N., Goswami, M. and Bhonde, R.R. 2005. Development of cell culture from liver and kidney of Indian major carp, *Labeo rohita* (Hamilton). *Indian Journal of Fisheries*. 52(3): 373-376.

Lakra, W.S. and Bhonde, R.R. 1996. Development of primary cell culture from the caudal fin of an Indian Major carp, *Labeo rohita* (Ham.). *Asian Fishery Science*. 9: 149-52.

Lakra, W.S., Bhonde, R.R., Sivakumar, N. and Ayyappan, S. 2006. A new fibroblast like cell line from the fry of golden mahseer *Tor putitora* (Ham). *Aquaculture*. 253: 238-43.

Lakra, W.S. and Goswami, M. 2011. Development and characterization of a continuous cell line PSCF from *Puntius sophore*. *Journal of Fish Biology*. 78: 987-1001.

Lakra, W.S., Goswami, M., Swaminathan, T.R. and Rathore, G. 2010b. Development and characterization of two new cell lines from common carp, *Cyprinus carpio* (Linn.). *Biol. Res.*43 (4): 385-392.

Lakra, W.S., Goswami, M., Yadav, K., Gopalakrishnan, A., Patiyal, R.S. and Singh, M. 2011b. Development and characterization of two cell lines PDF and PDH from *Puntius denisonii* (Day 1865). *In Vitro Cellular and Developmental Biology - Animal*.47(2): 89-94.

Lakra, W.S., Raja Swaminathan, T. and Joy, K.P. 2011a. Development, characterization, conservation and storage of fish cell lines: a review. *Fish Physiology* and *Biochemistry*. 37: 1-20.

Lakra, W.S., Swaminathan, T.R., Rathore, G., Goswami, M., Yadav, K. and Kapoor, S. 2010a. Development and characterization of three new diploid cell lines from *Labeo rohita* (Ham.). *Biotechnology Progress*. 26: 1008-1013.

Miller, N.W. and Clem, L.W. 1984. Temperature-mediated processes in teleost immunity: differential effects of temperature on catfish *in vitro* antibody responses to thymus-dependent and thymus-independent antigens. *Journal immunology*. 133: 2356-2359.

Nathiga Nambi, K.S., Abdul Majeed, S., Sundar Raj, N., Taju, G., Madan, N., Vimal, S. and Sahul Hameed, A.S. 2012. *In vitro* white spot syndrome virus (WSSV) replication in explants of heart of freshwater crab, *Paratelphusa hydrodomous*. *Journal of Virological Methods*. 183:2, 186-195.

Olivier, G., Eaton, C.A. and Campbell. 1986. Interaction between *Aeromonas salmonicida* and periteoneal macrophages of brook trout (*Salvelinus fontinalis*), *Veterinary Immunology and Immunopathology*. 12: 223-234.

Parameswaran, V., Ahmed, V.P.I., Ravi Shukla, Bhonde, R.R. and Sahul Hameed, A.S. 2007a. Development and Characterization of Two New Cell Lines from Milkfish (*Chanos chanos*) and Grouper (*Epinephelus coioides*) for Virus Isolation. *Marine Biotechnology*. 9(2): 281-291.

Parameswaran, V., Rajesh Kumar, S., Ahmed, V.P.I. and Sahul Hameed, A.S. 2008. A fish nodavirus associated with mass mortality in hatchery-reared Asian Sea bass, *Lates calcarifer*. *Aquaculture*. 275: 366-369.

Parameswaran, V., Ravi Shukla, Bhonde, R.R. and Sahul Hameed, A.S. 2006a. Splenic cell line from sea bass, *Lates calcarifer*: establishment and characterization. *Aquaculture*. 261: 43-53.

Parameswaran, V., Ravi Shukla, Ramesh Bhonde. and Sahul Hameed. A.S. 2006b. Establishment of embryonic cell line from sea bass (*Lates calcarifer*) for virus isolation. *Journal of virological methods*. 137: 309-16.

Parameswaran, V., Ravi Shukla, Ramesh Bhonde. and Sahul Hameed. A.S. 2007b. Development of a pluripotent ES-like cell line from Asian sea bass (*Lates calcarifer*)-An oviparous stem cell line mimicking viviparous ES cells. *Marine Biotechnology*. 9: 766-75.

Pettitt, T.R., Rowley, A.F. and Secombes, C.J. 1989. Lipoxins are major lipoxygenase products of rainbow trout macrophages, *FEBS Letter*. 259: 168-170.

Prasanna, I., Lakra, W.S., Ogale, S.N. and Bhonde, R.R. 2000. Cell culture from fin explant of endangered Mahseer *Tor putitora* (Hamilton). *Current Science*. 79(1): 93-5.

Rao, K.S., Joseph, M.A., Shanker, K.M. and Mohan, C.V. 1997. Primary cell culture from explants of heart tissue of Indian Major carps, *Current Science*. 73: 374-75.

Sahul Hameed, A.S., Parameswaran, V., Ravi Shukla, Bright Singh, I.S., Thirunavukkarasu, A.R. and Bhonde, R.R. 2006. Establishment and characterization of India's first marine fish cell line (SISK) from the kidney of sea bass (*Lates calcarifer*). *Aquaculture*. 257: 92-103.

Sashikumar, A. and Desai, P.V. 2008. Development of primary cell culture from *Scylla serrata*. *Cytotechnology*. 56: 161-169.

Sathe, P.S., Basu, A., Mourya, D.T., Marathe, B.A., Gogate, S.S. and Banerjee, K. 1997. A cell line from the gill tissues of Indian cyprinoid *Labeo rohita*. *In Vitro Cellular and Developmental Biology-Animal*. 33: 425-427.

Sathe, P.S., Mourya, D.T., Basu, A., Gogate, S.S. and Banerjee, K. 1995. Establishment and characterization of a new fish cell line MG-3 from gills of Mrigal (*Cirrhinus mrigala*). *Indian Journal of Experimental Biology*. 35: 589-99.

Secombes, C.J. 1985. The in vitro formation of teleost multinucleate giant cells. *Journal of fish Diseases*. 8: 461-464.

Singh, I.S.B., Rosamma, P., Raveendranath, M. and Shanmugam, J. 1995. Development of primary cell cultures from kidney of freshwater fish *Heteropneustus fossilis*. *Indian Journal of Experimental Biology*. 33: 595-599.

Sobhana, K.S., George, K.C., Venkat Ravi, G., Gijo Ittoop. and Paulraj, R. 2008. A cell culture system developed from heart tissue of the greasy grouper, *Epinephelus tauvina* (Forsskal 1775) by enzymatic dissociation. *Indian Journal Fisheries*. 55(3): 251-256.

Sobhana, K.S., George, K.C., Venkat Ravi, G., Gijo Ittoop. and Paulraj, R. 2009. Development of a Cell Culture System from Gill Explants of the Grouper, *Epinephelus malabaricus* (Bloch and Shneider). *Asian Fisheries Science.* 22: 1-6.

Sovenyi, J.F. and Kusuda, R. 1987. Kinetics of *in vitro* phagocytosis by cells from head-kidney of common carp, *Cyprinus carpio. Fish Pathology.* 22: 83-92.

Sunil Kumar, G., Singh, I.B.S. and Philip, R. 2001. Development of cell culture system from the ovarian tissue of African catfish (*Clarias gariepinus*). *Aquaculture.* 194: 51-62.

Suthindhiran, K., Sarath Babu, V., Kannabiran, K., Ishaq Ahmed, V.P. and Sahul Hameed, A.S. 2011. Anti-fish nodaviral activity of furan-2-yl acetate extracted from marine *Streptomyces* spp. *Natural Product Research.* 25: 834-843.

Swaminathan, T.R., Lakra, W.S., Gopalakrishnan, A., Basheer, V.S., Khushwaha, B. and Sajeela, K.A. 2010. Development and characterization of a new epithelial cell line PSF from caudal fin of Green chromide, *Etroplus suratensis* (Bloch, 1790). *In Vitro Cellular and Developmental Biology Animal*. 46(8): 647-656.

Swaminathan, T.R., Lakra, W.S., Gopalakrishnan, A., Basheer, V.S., Kushwaha, B. and Sajeela, K.A. 2012. Development and characterization of a fibroblastic-like cell line from caudal fin of the red-line torpedo, *Puntius denisonii* (Day) (Teleostei: Cyprinidae). *Aquaculture Research.* 43(4): 498-508.

Taju, G., Abdul Majeed, S., Nambi, K.S.N., Sarath Babu, V., Vimal, S., Kamatchiammal, S. and Sahul Hameed, A.S. 2012. Comparison of in vitro and in vivo acute toxicity assays in *Etroplus suratensis* (Bloch, 1790) and its three cell lines in relation to tannery effluent. *Chemosphere.* 87:1, 55-61.

Trust, T.J., Key, W.W. and Ishiguro, E.E. 1983. Cell surface hydrophobicity and macrophage association of *Aeromonas salmonicida. Current Microbiology.* 9: 315-318.

Uma, A., Prabhakar, T.G., Koteeswaran, A. and Ravikumar, G. 2002. Establishment of primary cell culture from hepatopancreas of *Penaeus monodon* for the study of whitespot syndrome virus (WSSV). Asian Fish. Science. 15: 365-370.

Wolf, K. and Quimby, M.C. 1962. Established eurythermic line of fish cells *in vitro*. Science, 135(3508): 1065 – 1066.

Chapter 15

Toll Like Receptors in Fishes: Identification and Functional Analysis

G. Dhinakar Raj

*Department of Animal Biotechnology,
Madras Veterinary College, Chennai – 600 007*

The greatest achievement of modern medicine is the prevention of infectious diseases through immunization. However, with years passing by after the first successful vaccination by Edward Jenner there had been considerable challenges in improving the efficacy of existing vaccines as well as efforts to find out targets for developing new and improved vaccines. Irrespective of the vaccine used or the invading pathogenic microorganisms the immune responses can be classified as "innate" and "adaptive". The highly sophisticated system of the adaptive immunity and its memory to recognize the 'non-self' with greater affinity is observed only in vertebrates. In contrast, the innate immune system has been shown to be phylogenetically conserved and is present in almost all multicellular organisms. The mechanism by which this happens in different species was not known until quite recently. The innate immune surveillance relies in part by recognizing conserved molecules unique to some classes of potential pathogens like lipopolysaccharide (LPS) and lipoteichoic acid (LTA) found on the surface of gram negative and positive bacteria respectively. These molecules have been collectively called as pathogen-associated molecular patterns (PAMPs). In this connection a cell surface molecule namely the TOLL receptor that can recognize these PAMPs had gained attention in the recent years. This was originally identified in Drosophila to be essential for the dorso-ventral pattern in the developing embryos and fruit-flies with mutant Toll receptor were found to be highly susceptible to fungal infections. Recently a family of type I transmembrane pattern-recognition receptors (PRR) that are structurally related to drosophila Toll namely the Toll-like receptors (TLR) have been identified in most of the species that endow the host with an efficient, non self-reactive means to detect the invading pathogens.

TLR: Their Structure, Classification and Signaling

TLRs represent type I transmembrane receptors defined by the presence of leucine-rich repeats (LRRs) in their extracellular domain which are responsible for both microbial recognition and receptor dimerisation and by the presence of a Toll/interleukin-1 (IL-1) receptor domain (TIR domain) in the C-terminal, cytosolic part of the protein that initiates signal transduction.

Figure 15.1: Goat TLR 1 Protein Domain Predicted by SMART Analysis

The TIR domain is an evolutionary ancient protein–protein interaction domain that occurs in a large group of host defense associated proteins from diverse species, including vertebrates, invertebrates and plants. TIR-domain containing proteins in the human genome comprise ten members of the TLR family, eight members of the IL-1/IL-18 receptor group (IL-1R/IL-18R) and five identified or predicted cytosolic proteins that function as adaptor proteins connecting the Toll-like or interleukin receptors with the downstream signaling pathways.

The first adaptor protein that was described is encoded by the myeloid differentiation primary response gene MyD88, which has been implicated in signaling by all of the human TLRs with one possible exception (TLR3). In addition to its TIR domain, MyD88 contains a death domain that associates with the death domain of IL-1R-associated kinase (IRAK). Subsequent autophosphorylation of IRAK enables its interaction with tumour-necrosis-factor-associated factor-6 (TRAF-6), ultimately leading through two diverging signaling pathways to the activation of nuclear factor kB (NF-kB) and mitogen-activated protein (MAP) kinase. IN addition, two novel adaptors called MAL (MyD88-adaptor-like protein, or TIRAP) and TRIF (TIR-domain containing adaptor inducing INF-β, or TICAM-1) have been described, which are involved in signaling by distinct TLRs. Furthermore, the human genome contains two more potential adaptor proteins, TRAM (TRIF-related adaptor protein) and SARM (sterile α and HEAT-Armadillo motifs), the differential use of which might further explain how distinct pathways are activated by TLRs during host defense.

The constituent components of the TLR pathway can be divided into three distinct groups:

- ☆ TLRs, the molecules responsible for pathogen recognition,
- ☆ The intracellular signaling components that relay the immune signal to
- ☆ The downstream effector molecules that execute the host response.

The TLR families are known to consist of 10 members (TLR1-TLR10) however, there number is still increasing. They are widely expressed in many cell types, including non-haematopoietic epithelial and endothelial cells; although most cell types express only a select subset of these receptors.

The TLR family is generally subdivided based on the genomic structure into five subfamilies: the TLR3, TLR4, TLR5, TLR2 (composed of TLR1, TLR2, TLR6 and TLR10) and TLR9 (composed of TLR7, TLR8 and TLR9).

The signaling starts with the TIR domain interacting with adaptor proteins like MyD88, TIRAP, TICAM1 and TICAM2 which leads to the specific activation of specific transcription factors namely interferon regulatory factor (IRF3) or Nuclear factor-??beta or AP-1. These transcription factors translocate to the nucleus to regulate the expression of type I interferon or pro-inflammatory cytokines.

Mouse TLR stimulation results in signal through one of five adaptor proteins (MyD88, TIRAP/MAL, TICAM-1, TICAM-2 and SARM). All TLRs with the exception of TLR 3, can signal through the adaptor molecule MyD88. MyD88 recruits members of the interleukin-1 receptor-associated kinase (IRAK) family which in turn activate the key ubiquitin E3 ligase, tumour necrosis factor receptor-associated factors (TRAFs) and transforming growth factor-β (TGF- β)-activating kinase (TAK1), leading to the

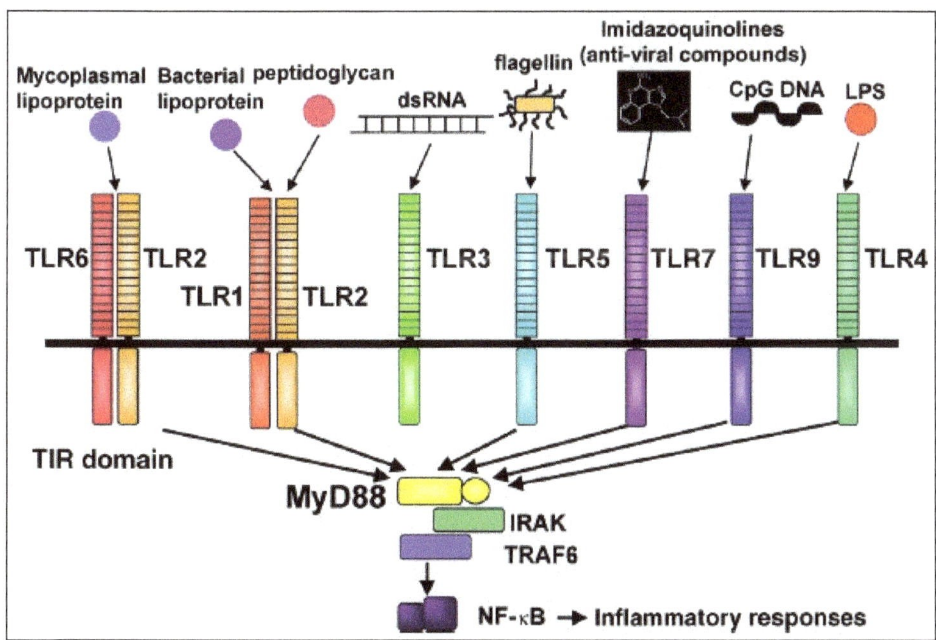

Figure 15.2: TLRs Recognize Molecular Pattern Associated with Bacterial Pathogens. Triacylated Lipoprotein for TLR1; Peptidoglycan for TLR2; double-stranded RNA for TLR3; lipopolysaccharide (LPS) for TLR4; flagellin for TLR5; diacylated lipoprotein for TLR6; imidazoquinoline and its derivative R-848, for TLR7; and bacterial unmethylated CpG DNA for TLR9. MyD88 associates with the TIR domain of TLRs and transduces signals to induce immune responses.

activation of the transcription factor NF-k. The activated NF-kB is translocated to the nucleus and induces expression of inflammatory cytokine genes such as IL-1β and IL-6. In contrast, TLR3 recruits the adaptor molecule TICAM-1, leading to activation of interferon (IFN) regulatory factor-3 (IRF- 3) and induces an antiviral response through expression of type I IFN and IFN-inducible genes.

TLR and Induction of Antimicrobial Activity

TLR induction had been shown to be involved in the development of antimicrobial activity. One such response on activation of TLR2 had been shown to result nitric oxide-dependent and –independent killing of intracellular pathogen like *M. tuberculosis.* Recent reports also indicate that the TLRs are likely to mediate the secretion of antimicrobial peptides like defensins to result in direct killing of microbes at epithelial surface. Induction of apoptosis of infected macrophages may limit the spread of pathogens and in this connection TLR2 mediated activation of the adaptor protein MYD88 has been shown activate the apoptotic pathway involving the caspase 8.

TLR in Fish Species

To date, 17 different TLRs have been identified in more than a dozen different fish species. Numerous studies revealed that specific piscine TLRs share functional properties with their mammalian counterparts. Nevertheless, remarkable distinct features of teleostean Toll-like receptor cascades have been discovered. A soluble TLR5 factor in rainbow trout for example might amplify danger signaling of membrane-

Table 15.1: Comparison of TLR Repertoire among different Species

TLRs	*Human*	*Mouse*	*Chicken*	*Zebra fish*	*Fugu*	*Lamprey*
TLR 1	+	+	+	+	+	–
TLR 2	+	+	+	+	+	–
TLR 3	+	+	+	+	+	+
TLR 4	+	+	+	+	–	–
TLR 5	+	+	+	+	+	+
TLR 6	+	+	Psd	–	–	–
TLR 7	+	+	+	+	+	+
TLR 8	+	+	+	+	+	+
TLR 9	+	+	–	+	+	–
TLR 10	+	Psd	–	–	–	–
TLR 12 (TLR 11)	–	+	–	–	+	–
TLR 13	–	+	–	–	–	–
TLR 14 (TLR15)	–	–	+	+	+	+
TLR 21	–	–	+	+	+	+
TLR 22 (TLR23)	–	–	–	+	+	+
TLR 24	–	–	–	–	–	+

bound TLR5 in a positive feed loop. Piscine TLR3 detects viral and additionally bacterial molecular patterns in contrast to mammalian TLR3. Regarding TLR4, the functional spectrum of this teleostean receptor is also different from its mammalian orthologue. While signaling quite similar as the mammalian counterpart in some fish species, it may down regulate TLR activation in others or was even lost during evolution. The orthologues of human TLR6 and TLR10 are also absent in teleosts. Some piscine TLRs are encoded by duplicated genes, for example salmonid TLR22.

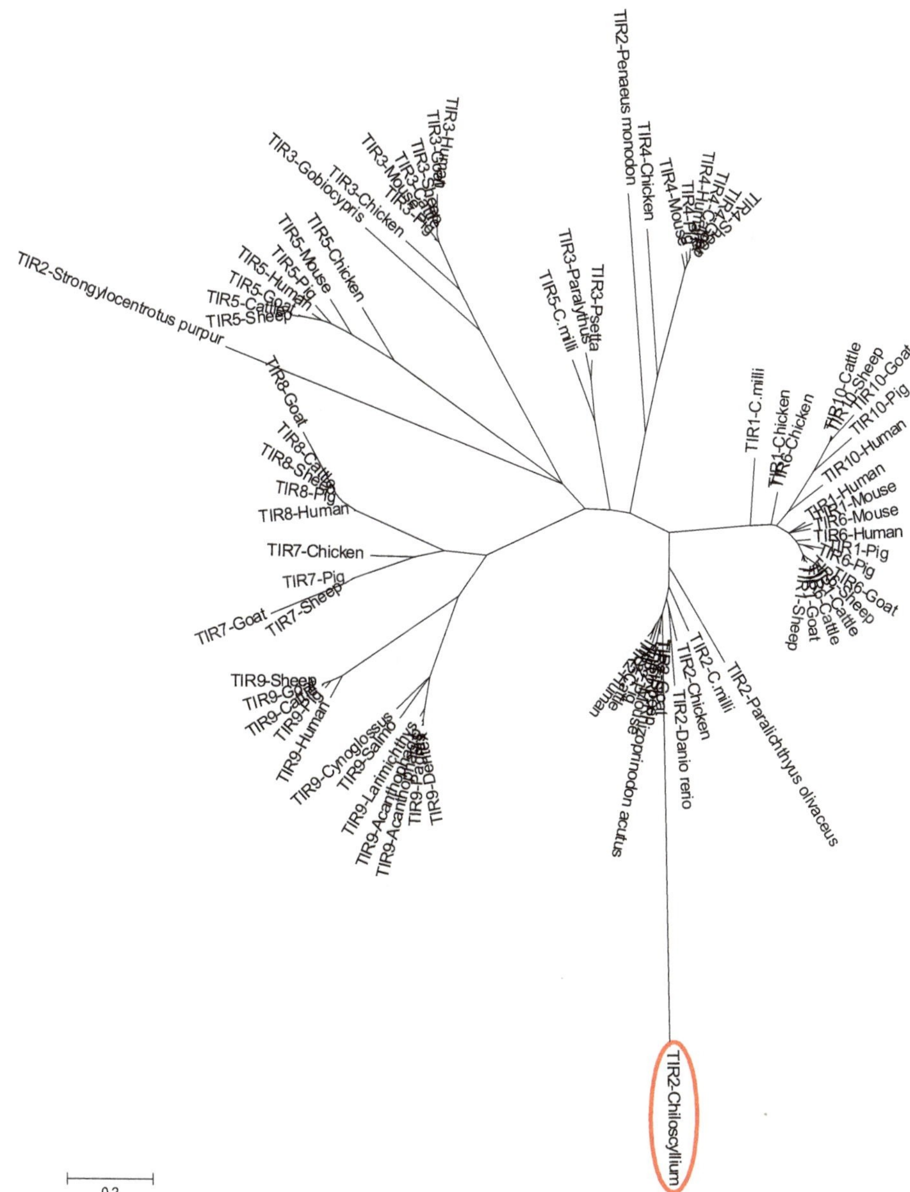

TLR22 is found in several fish species but only as a non-functional pseudogene in man. The expression profile of only some TLRs and their ligands have been characterized to some extent in zebra fish (*Danio rerio*), puffer fish (*Takifugu rubripes*), catfish (*Ictalurus punctatus*), rainbow trout and flounder. Some information is also available from the shrimp *Penaeus monodon* and *Litopenaeus vannamei.*

TLRs in Sharks

We have shown the presence of TLRs in sharks (see below). The TIR of TLR 2 showed highest up regulation (12.8 folds) in the skin following inoculation of a TLR ligand pool intra peritoneally to *Chyloscillium* species of sharks.

TLR 22 in Fish

Teleost fish TLR 22 have not been found from mammals so far. TLR 22 is not included in any clade of human TLRs, and they likely originated around the Cambrian period; this indicates that the human ancestor possessed TLR 22 gene. The importance of TLR22 in teleosts is shown by its activity against RNA viruses. TLR22 is located on the cell surface membrane, TLR22 prefers to recognize long dsRNA (< 1 kbp) and it is the only TLR that can recognize nucleic acids on the cell surface and transmit signals to induce cytokines.

TLR22 is widely conserved among teleosts and amphibians, but extensive genome projects failed to reveal the presence of the TLR22 gene in avian or mammalian genomes. Therefore, it seems likely that TLR22 is required for vertebrates that live in the water. TLR22 is ubiquitously expressed in puffer fish tissues, but tissue-specific expression, with strong expression in the head and kidney, mild expression in the trunk, spleen, and gill, and undetectable expression in the intestine, liver, brain, and skin, has been reported for the Japanese founder (*Paralichthys olivaceus*). This finding illustrates that the expression pattern is different among different teleosts. Interestingly, both the puffer fish and Japanese flounder TLR22 genes are upregulated by stimulation with polyIC, which is a synthetic analog of double-strand RNA. Therefore, the TLR22 function seems to be conserved among teleosts. Teleosts possess two viral RNA-recognizing TLRs: TLR3 and TLR22. Double-strand RNA derived from RNA virus is recognized by TLR22 on the cell surface and simultaneously by TLR3 in the early endosome.

TLR and its Application in Fishes: Future Prospect

The role of TLR in linking innate and adaptive immunity has created new avenues in improving the efficiency of vaccines by including TLR ligands. Finding out suitable adjuvants that enhance the immune response against a particular target is also major and recurrent issue. There is scanty literature regarding the use of TLR in modulating the immune response of fish. By including suitable ligands it is possible to induce low-level stimulation of TLRs and to skew the response towards Th2 or Th1-type response. It is also possible to facilitate the migration of dendritic cells (DC) to regional lymph nodes upon TLR stimulation which happens through the upregulation of the chemokine receptor CCR7. The unmethylated CpG dinucleotides of bacterial DNA that serve as a main stimulus for TLR9 are very much useful as vaccine adjuvants.

Synthetic low molecular compounds namely imidzoquinolines have been shown to possess antiviral and antitumour activity in several animal models through stimulation of TLR7. S-27609 a synthetic ligand for TLR7 had been tried in Atlantic salmon to induce both an "innate" IFN response and IFN-? response there by resulting in better antiviral immune response. In addition the presence of soluble forms of TLR like the sTLR5 has been shown to serve as a good adjuvant for flagellin-mediated activation of human immune system.

Chapter 16

Algal Culture Techniques for Larval Rearing of Ornamental Fishes

M.K. Anil, Rani Mary George, S. Jasmine and B. Santhosh

Research Centre of the Central Marine Fisheries Research Institute, Vizhinjam, Kerala – 695 221

Unicellular marine algae are widely used as live feed in the hatchery production of commercially valuable fish and shellfish. Algal production unit is a vital part of any hatchery. Shrimp, oyster, clam and mussel larvae feed directly on microalgae. Other important live feed such as rotifers and *Artemia* also feed on algae and are cultured for use as feed for larval fish and shrimps. In CMFRI Fish Hatchery Unit, larval rearing is done in green water systems by adding algae to the water containing fish or shrimp to improve the water quality and to give nutrition to other zooplanktonic live feeds in the tank.

Marine algae are plants which contain chlorophyll and trap the energy from light and convert nutrients and carbon dioxide dissolved in the sea water into organic growth. In the laboratory or hatchery, a collection of algal cells, which are growing by multiplication, is known as stock culture of phytoplankters.

Of the very many types of algae which live in the sea, only a few can be cultured, and only certain types will give good growth when fed to fish larvae, oysters, clams, rotifers, brine shrimps, etc.

Table 16.1: Commonly Used Algae for Culture

Species of Algae	*Size in Microns*	*Species of Algae*	*Size in Microns*
Chlorella marina	8-10	*Nannochloropsis occulata*	2
Chaetoceros calcitrans	2.5	*Isochrysis galbana*	5
Chaetoceros gracilis	6	*Tetraselmis suecica*	8.5
Skeletonema costatum	6	*Monochrysis lutheri*	4
Thalassiosira pseudonana	5.5	*Dunaliella trtiolecta*	6.5

Flagellates can swim by the action of one or more flagellae (*Nannochloris, Dunaliella, Tetraselmis, Isochrysis, Monochrysis*) and the diatoms (*Skeletonema, Chaetoceros, Thalassiosira*) which have an outer shell composed of silica can remain buoyant. Individual species of algae can be used as feed, but feeding mixed diets, consisting of at least one type of diatom and one type of flagellate, nearly always give much better growth.

For any successful hatchery operation, production of healthy monocultures of selected algal species is a necessity. Algae are harvested from large vessels whose production is reliant on a reservoir of small stock cultures. It is crucial to maintain this stock free of contaminants and excessive bacteria. Contamination of algal cultures may occur *via* the seawater supply, air supply and cross-contamination from nearby algal cultures. For this reason, stock cultures are maintained in a separate unit from the larger culture containers. Cleanliness and careful transfer techniques cannot be overemphasized in maintaining a functional and healthy algal culture operation.

Requirements for Algal Cultures

Several factors are essential for algal growth. Attempting to achieve as constant culture conditions as possible will favour optimal algal growth, which will furthermore affect their biochemical composition.

Salinity

Filtered seawater of required salinity is the most basic requirement for the marine microalgal culture. Generally all marine algae used in the hatchery can be cultured in the normal seawater salinity of 35 ppt. Salinities of between 25 and 30 ppt. are generally best for the culture of flagellates, and between 20 and 25 ppt for the culture of diatoms. The salinity can be obtained by diluting sea water with tap water. Salinity can be measured with a refractometer or hydrometer.

Nutrients

For rearing high concentrations of algae, type and the amount of nutrients added is critical. Nutrients are the inorganic salts required for plant growth. For algae culture, it is convenient to make up strong standard solutions which, in appropriate dilutions in sea water, provide the culture medium. There are several types of nutrient media which may be used; they can either be made using chemicals, or can be purchased ready-made from aquaculture companies. Recipes for the nutrient salt solutions, used for culture of algae at fish hatchery of Central Marine Fisheries Research Institute, Vizhinjam (Conway medium, F/2) are provided in Appendix I and II.

Light and Temperature

This is normally provided by fluorescent lamps or tube lights. Increasing the light intensity usually means better growth and faster division of algal cells and therefore, the production of more feed. It is important to make the most efficient use of the artificial light as lamps which also generate heat and may make the culture too hot. Cultures of algae can also be grown outdoors, using natural daylight, as done in our hatchery. Most types of algae grow well at temperatures from 22 to 28°C. Very Low temperatures will not kill the algae, but will reduce their growth rate. Above 29°C, most types of algae will die. If necessary, cultures can be cooled by a flow of cold water over the surface of the culture vessel or by controlling the air temperature with air conditioning units.

Other Factors

Daily shaking of stock culture or bubbling of air through mass culture will make the necessary light and nutrients available to all the cells and prevent algae from settling. Algae cultures are usually mixed by bubbling air through them. This can be supplied from a compressor or via an air blower, and will also act as a carrier gas for carbon dioxide. Providing the algae with extra carbon, in the form of the gas carbon dioxide (CO_2) will give much faster growth. CO_2 is supplied from compressed gas cylinders (Not an essential in small scale operations), and only a very little is needed (less than one percent) in the air supplied to the culture. The CO_2 should be passed through a flow meter to ensure that the amount used will keep the pH of the culture between 7.8 and 8.0. The pH can be checked with indicator papers, which change colour with a change in pH, or with a pH meter. Both the air and the CO_2 should be filtered through an in-line filter unit of 0.3 microns to 0.5 microns before entering the culture, as this helps to prevent other, possibly contaminating, organisms from getting into the cultures.

Algal Growth

Growth of unicellular algae is by simple cell division, *i.e.*, a single cell divides to form two cells, which then divide to form four cells, etc. Under normal conditions, an algal culture goes through three phases of growth: lag, exponential and stationary phases (Figure 16.1). The lag phase occurs when the culture is started and little increase in cell density is observed. In healthy cultures this period is quite short. In the exponential phase cell division occurs rapidly and cell density increases geometrically. Growth is limited only by the time required for cell division in this phase. In the stationary phase, the rate of growth (cell division) declines because some factors, such as nutrients or light, has become limiting and cell density remains relatively constant. During these different phases of growth the biochemical content of the algae differs. It has been shown that higher energy levels are found in the stationary phase for most species. In a hatchery situation, a balance between high cell density and optimal energy content is strived for and harvest them at the beginning of the stationary phase, when energy content peaks.

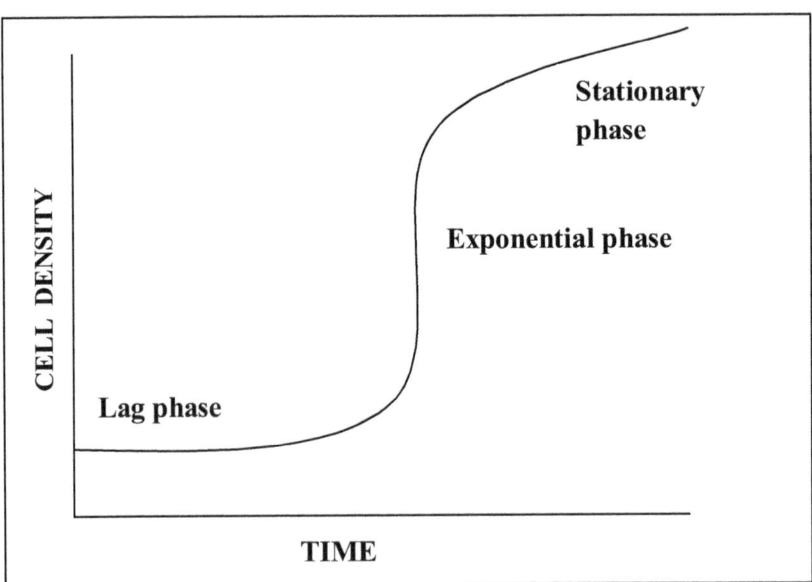

Figure 16.1: Theoretical Growth Curve of Typical Algal Culture Showing Lag, Exponential and Stationary Phase (take from Bourne, Hodgson and Whyte, 1989)

Culture Techniques

The sea water used for culture must be clean or unwanted types of algae and other contaminants, which may feed on or compete with the algae, will grow in the cultures. Small amounts (up to about 4 litres) of sea water can be autoclaved (sterilized by steam at a high pressure in a pressure cooker), or pasteurized (heated to 80°C for 1-2 hours, cooling to room temperature for at least 18 hours, then reheated to 80°C for a further 1-2 hours) or boiled. A container of suitable material *e.g.* borosilicate glass (Pyrex)) can be used.

Larger volumes of sea water can be cleansed by filtration or by chlorination. There is a wide range of suitable equipment commercially available for this purpose, which includes cartridge filters, filter assemblies using diatomaceous earth as an aid to efficient filtration, and swimming pool type rapid sand filters. The main consideration is that the equipment chosen can cope with the volume of sea water that is needed. Filtration to remove particles greater than 2 microns is essential, and the removal of particles larger than half a micron is desirable. A large quantity of sea water can also be cleansed by passing it slowly through a ultra-violet light sterilizing unit following filtration to remove particles greater than 2 microns in diameter.

The following sections provide the protocols used at the CMFRI hatchery for culture of algae from stock to 20 litres drinking water cans used for small scale hatcheries. In order to ensure a contaminant free culture, sterile microbiological techniques are used during sub-culturing or startup of solutions. These techniques rely on the flaming of all equipment prior to use, and the continuous working by the flame during addition of nutrients, transfer of cultures, etc. At the hatchery, a spirit lamp or a Bunsen burner used as a flame source. Prior to opening of any culture flasks or nutrient bottles, the flame is lighted on, placed on the workbench, and left on throughout the inoculation process. In this way, the area is continuously flamed and "sterilized" when in use.

Stock Cultures

All algal culture system requires a set of "stock" cultures; these are usually of about 125–250 ml in volume, and provide the reservoir of algal cells from which to start the larger-scale cultures used for feeding. Several centres, which specialize in the culture of algae can provide inoculum for stock cultures. However, it is also possible to isolate algal species from a specific body of water, and attempt to rear them under controlled conditions for feed. The isolation of singles cells of a species from natural live phytoplankton samples can be done using a capillary pipette and/ or via a series of dilutions; this allows the separation of a selected species into a culture chamber with nutrient media. Thereafter, algae are cultured as detailed below. At our hatchery, master algal cultures are received from CMFRI algal culture laboratories.

The procedure for starting algal cultures and growing them to high density in 125-250 ml flask is described below. The following section describes the culturing of 1000 ml - 4-litre and 20-litre cultures. Prior to reception of purchased cultures, 250 ml Erlenmeyer flasks are cleaned and prepared with adequate salinity seawater (see Protocol–1).

Erlenmeyer flasks (250 ml) are autoclaved with seawater, for sterilizing of both flasks and seawater. Full strength salinity (35 ppt) is used for all cultures, except for the diatoms, *Chaetoceros* species and *T. pseudonana*, which grow best in reduced salinity seawater (25 ppt). Once the autoclaving process is completed, cooled flasks are ready for inoculation. Finely filtered (0.2 to 1 ìm) and autoclaved seawater should be adequate.

Preparation of Culture Water

Seawater can be collected from any clean coastal belt. Water with a high load of particulate matter especially polluted locations should be avoided. All containers for carrying or storing seawater must be clean. It is advisable to use dedicated containers as you can be sure that they have not been used to store harmful chemicals. Collected seawater should be stored for a few days before being used for algal culture. Seawater must be filtered through a filter bag or pumping through a catridge filter of 5 and 2 microns.(~ 5ìm) and sterilized before being used in cultures.

In small scale commercial hatcheries sea water can be stored in large PVC tanks of 5-10 tons or in cement tanks of 10 to 100 tons. This water is then filtered by passing through sand filters or high pressure sand filters and then passed through a series of cartridge filters of 5, 2 and 0.2 micron. Filtered water can also be passed through a UV filter. Small volumes can be heated in a microwave oven. Some literature suggests autoclaving but this is only needed for specialized work with very sensitive species of algae. If water is heated twice, at 24 hour intervals, it will definitely be clean enough for general algal culture. In areas with strong sunlight, glass or PET plastic containers of seawater in direct sunlight will receive UV light and radiant heat. If this water reaches ~50°C it can be used for algal culture when it has cooled. After sterilization it is best to allow water to cool slowly by standing for at least 24 hours. This will allow some CO_2 to diffuse from the atmosphere into the water.

For larger volumes of water, heating is inconvenient and uneconomical. However, chlorination/dechlorination is an effective and easier way to sterilize. For this technique, bleach (sodium hypochlorite) is used to kill all life in the culture water. Sodium thiosulphate is then used to deactivate the remaining bleach. Liquid pool chlorine is the most appropriate source of bleach as it contains a high concentration of active chlorine (125-150g/L) and has no added surfactants or perfumes. Sodium thiosulphate should be prepared as a 100 g/L stock solution. To sterilise algal culture water, add liquid pool chlorine at the rate of 1.5 ml per 10L of water. It is best to place the culture water in the culture vessel and then add the aeration tubing so that both are sterilized along with the water. Cover the vessel and leave for 24 hours without aeration. After 24 hours, add the thiosulphate solution at the same rate (*i.e.*, 1.5 ml per 10L) and aerate vigorously for an hour before adding nutrients and inoculating with algae. Chlorotex or any chlorine test kit can be used ensure that the medium is free of any residual chlorine.

Protocol–1

Preparation of Culture Flasks (125 ml–250 ml)

1. Wash glass flasks, pipettes in 10 per cent HCl (hydrochloric acid) bath.
2. Rinse 3 times with fresh water and do a final rinse with distilled water and let dry.
3. Fill flask with appropriate volume (100 ml in 250 ml flask) of filtered seawater using a graduated cylinder.
4. Close flask loosely with cotton plug and aluminium foil. Cotton plugs are made with cheesecloth material tied around absorbent cotton.
5. Place flasks in autoclave/pressure cooker and or boil for 2 minutes.
6. Wrap all pipettes in aluminium foil and sterilize in hot air oven or in autoclave/ pressure cooker or boil in a water bath for 5 minutes.
7. Cool the sterilized flask containing seawater overnight and add nutrients (A,B and C) solution (See Appendix I) using sterile pipettes in front of a flame
8. Unplug the flask in front of a flame and add 50 ml of inoculum or starter culture into the prepared medium in the flask, plug the flask and mix well and place under the fluorescent light.

Figure 16.2: Starter Culture.

9. Mix the culture daily by gentle shaking and the culture will be ready within 5 to 7 days and can be used to inoculate 1000 ml flasks.

Protocol-2

Preparation of Culture Flasks (1000 ml-4000 ml)

1. Prepare the flask as in protocol 1 and fill the flask with filtered seawater to half the volume.
2. Plug the flask with cotton and boil the flask for 10 minutes and cool overnight.
3. Add the nutrient solution under aseptic condition and inoculate the medium from a 250 ml stock and place it under fluorescent light or in a room or veranda getting sufficient sunlight preferably facing north or south to avoid direct sun light which may heat up the culture.

Figure 16.3: Flask Culture.

4. Mix the culture daily by gentle shaking and the culture will be ready within 5 to 7 days and can be used to inoculate 20 l pet jars/drinking water jars.

Figure 16.4: Algal Culturing at Room Temperature.

Protocol-3

Preparation of Culture Jars (20 lit.)/ Polythene Bags (20-30 Lit)

1. Wash 20 litre jars and aeration tube and stones with 10 percent HCl (hydrochloric acid).
2. New polythene bags whoch are made using heat process are sterile inside and can be directly used in place of jar. These bags are hung with support.
3. Rinse 3 times with fresh water and do a final rinse with distilled water.
4. Fill jar/bag with appropriate volume (15-30 l) of filtered seawater from a cartridge filter set up of 2 micron or less or through a filter bag.
5. Put the aeration stone and its pipe line; close the jar with cotton plug.
6. Add 1.5 ml of sodium hypochlorite solution per 10 l of seawater in the jar. (10-12.5 per cent active chlorine), plug the vessel and leave for 24 hours without aeration.
7. After 24 hours add the sodium thiosulphate solution at the same rate (*i.e.*, 1.5ml per 10 l) and aerate vigorously for an hour before inoculating with algae.
8. Take 1or 2 ml of the water and check for any residual chlorine using chlorotex reagent, once standardized this step can be avoided.
9. Add the nutrient solution and inoculate the medium from a 2-3 l of culture and place it under fluorescent light or in a room or verandas getting sufficient sunlight preferably facing north or south to avoid direct sun light.

Figure 16.5: Chlorination of Seawater.

Figure 16.6: Outdoor Algal Culture (20 lit.)

10. Culture will be ready for use in 3-7 days depending on the strength and quantity of inoculums used, light and prevailing temperature.

Monitoring Algal Populations

A regular check of microalgal cultures is essential to prevent crashes and to keep high quality standards. The main parameters to be monitored are: colour, density, pH and contaminant levels. As an example, a change in colour to opaque grey and a pH level lower than 7.5 may indicate a high degree of bacterial contamination. A lighter colour than normal may reveal insufficient nutrients or poor lighting. Mass cultures are normally checked at naked eye by experienced staff and strict controls are usually restricted to pure strains and small vessels. The algal culture growth can be monitored daily by counting the number of cells per ml with a haemocytometer (Figure 16.7).

Determination of Algal Cell Density

The apparatus used for counting cells is a haemocytometer with an improved Neubauer ruling. Before counting, both the cover slip and chamber must be rinsed clean and dried. The face of the counting chamber is composed of two gridded surfaces

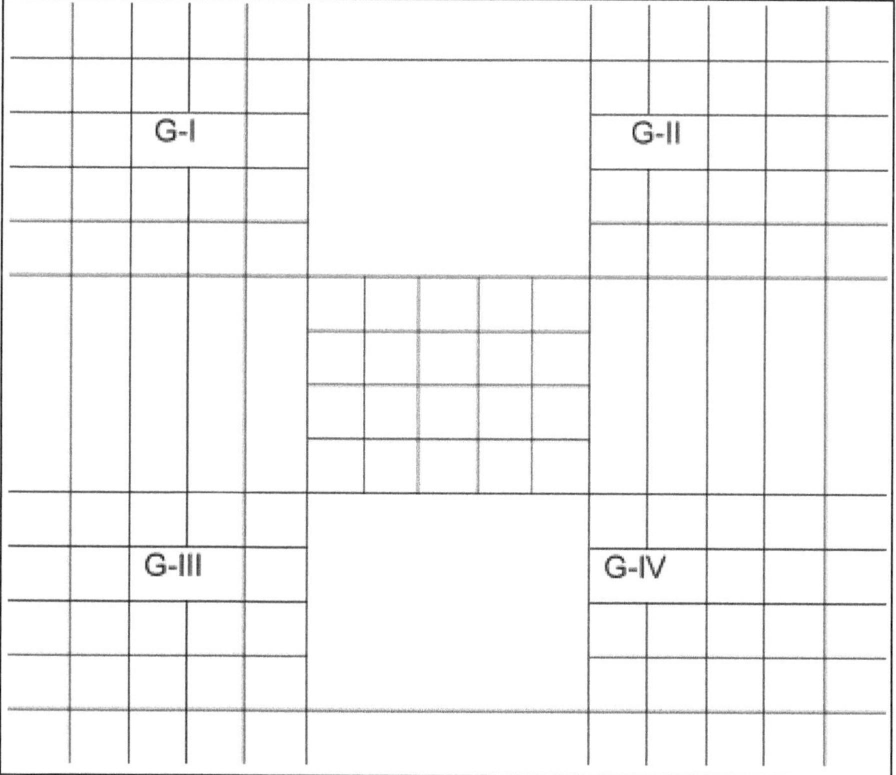

Figure 16.7: Surface View of N.C. Chamber (Haemocytometer) Showing Grid Areas (G = Grid)

separated by canals. The cover slip is placed on the support bars along the canals and a drop of homogeneously mixed algae suspension is delivered from a Pasteur pipette by touching the pipette tip to the edge of the cover slip where it hangs over the V-shaped loading port. Slight pressure will cause the algal suspension to flow evenly across the surface, but not into the canals or on top of the cover slip.

A small drop of 5 to 10 per cent formalin mixed into the sample is sufficient to immobilize cells for counting. Each half of the haemocytometer contains nine large grids. Only those algal cells which fall within the four large corner grids are counted. Each large corner grid is further subdivided into 16 small squares. Moving systematically back and forth across the squares, a minimum of 200 algal cells are counted in as many grids as necessary. To determine the algal cell density (number of algal cells per milli litre) in the suspension, the number of algal cells counted is divided by the large corner grid area covered, multiplied by 10,000. For example, if 300 algal cells were counted in 1.5 large corner grids (or 24 small squares), the cell density is 300 algal cells/1.5 corner grids x 10,000 = 2×10^{6} cells per ml.

References

Becker, E.W. (ED.). 1985. 'Production and Uses of Microalgae'. Schweizerbart'sche Verlagsbuchhandlung, Stuttgart, FRG. pp: 189.

De La Nove, J. and De Pauw, N. 1988. The potential of microalgal biotechnology: a review of production and uses of microalgae. *Biotechnol. Adv*. 6: 725-770.

Lain, I. and Ayala, F. 1990. Commercial mass culture techniques for producing microalgae. In: I. Akatsuka (Ed.), 'Introduction to Applied Phycology'. SPB Academic Publishing bv, The Hague, The Netherlands. pp: 447-477.

Richmond, A. (ED.). 1986. 'CRC Handbook of Microalgal Mass Culture'. CRC Press, Boca Raton, USA. pp: 536.

Shelef, G. and Soeder, C.J. (EDS.). 1980. 'Algae Biomass: Production and Use'. Elsevier/ North Holland Biomedical Press, Amsterdam. pp: 312.

Soeder, C.J. 1980. Massive cultivation of microalgae: results and prospects. *Hydrobiologia*. 72: 197-209.

Soeder, C.J. and Bonsack, R. (EDS.). 1978. Microalgae for food and feed. Ergebn. Limnol. 11:1-300.

Stein, J.R. (ED.). 1973. 'Handbook of Phycological Methods, Culture Methods and Growth Measurements'. University Press, Cambridge, UK. p. 448.

APPENDIX I
Conway or Walne's Medium

Solution A

Potassium nitrate	– 100 g
Sodium orthophosphate	– 20g
EDTA (Na)	– 45 g
Boric acid	– 33.4 g
Ferric Chloride	– 1.3 g
Manganese chloride	– 0.36 g
Distilled water	– 1 litre

Solution B

Zinc chloride	– 4.2 g
Cobalt chloride	– 4 g
Copper sulphate	– 4 g
Ammonium molybdate	– 1.8 g
Distilled water	1 litre

Solution C

Vitamin B1 (Thiamin)	– 200 mg in 100 ml distilled water
Vitamin B12 (Cyanocobalamine)	– 10 mg in 100 ml distilled water

Store A, B and C solutions in different reagent bottles. Add 1ml of A, 0.5ml of B and 0.1ml of C to one litre of filtered and sterilized seawater.

APPENDIX II
Guillard's f/2 Medium for Culturing Unicellular Algae

Preparing Media

These instructions are for making 50 aliquots of concentrated medium, each of 20 ml, which are to be frozen. Each 20 ml will contain f/2 nutrients for 2 L of culture. The quantities should be adjusted if larger cultures or more than 50 aliquots are needed. For example, to make 50 aliquots with each aliquot containing nutrients for 20 L of f/2 media, use 10 time the amounts given below.

1. Trace Elements

The first step is to prepare a standard stock solution of concentrated trace elements. Only a portion of this is used for each batch of f/2, the remainder can be frozen for later use.

*Trace Element Weigh Out**	*Volume of Water**
$CuSO_4.5H_2O$ 100 mg	100 ml
$ZnSO_4.7H_2O$ 220 mg	100 ml
$CoCl_2.6H_2O$ 100 mg	100 ml
$MnCl_2.4H_2O$ 1800 mg	100 ml
$Na_2MoO_4.H_2O$ 60 mg	100 ml

* If this amount cannot be weighed exactly, weigh an amount close to this and calculate the volume of water required to end up with the same concentration. *e.g.*, if 112 mg of $CuSO_4.5H_2O$ is weighed, add it to 112 ml of water. There are now five solutions, each of a trace element. Take exactly 50 ml of each solution and pool them in a 500 ml cylinder or volumetric flask and make the volume up to 500ml with water. The pooled solution of trace elements is more than needed to prepare 50 x 20 ml ice blocks, each for 2 L of culture. Only 10 ml is required. Keep 10 ml of this trace element solution. Freeze 10 ml aliquots for later use.

2. Vitamins

Prepare standard stock solutions of concentrated vitamins. As for the trace elements, most of this can be frozen for later use.

*Vitamin Weigh Out**	*Volume of Water**
Thiamine HCl 500 mg	50 ml
Biotin 50 mg	1000 ml
B12 50 mg	1000 ml

* As above, but 50 ml of Thiamine HCl is needed do not weigh out less than 500 mg.

Take 50 ml of each of the three vitamin solutions and pool them in 500 ml of water. Keep 10 mls of this vitamin solution. Freeze 10 ml aliquots for later use.

3. Silicate (Use for Culturing Diatoms Only)

Silicate is only required for growth of diatoms. It can be omitted from f/2 unless diatoms are grown. Therefore it is convenient to keep silicate separate from the other components of f/2.

Disolve 3.0 g in 100 ml of water.

Distribute 2 ml aliquots into vials or keep as a liquid, using 1 ml per litre of culture medium as required.

4. Major Chemicals

Weigh exactly each of the following and keep them in separate labelled vials.

$NaNO_3$	8.3g
$NaH_2PO_4.H_2O$	0.5g
$FeCl_3.6H_2O$	0.315g * caution, hygroscopic.
EDTA	0.436g

5. Preparation of 20ml aliquots

To about 500 ml of distilled or deionised water,

Add 0.436 g of the EDTA. Stir to dissolve. N.B. EDTA must be added first.

Add 0.315 g of the iron chloride. Dissolve.

Add 0.5 g of the phosphate salt

Add 8.3 g of the nitrate salt

Add 10 ml of the pooled trace metal solution

Add 10 ml of the pooled vitamin solution

Pour 20 ml aliquots of this solution into snap top vials, seal and freeze. These will keep almost indefinitely. For culturing diatoms, do not add silicate to the frozen aliquots. Add it directly to culture media along with frozen aliquot.

Chapter 17

Cryopreservation Techniques

T. Francis

Department of Fisheries Biology and Capture Fisheries, Fisheries College and Research Institute, Thoothukudi – 628 008, T.N.

Cryopreservation is a branch of cryobiology, which relates to the long-term preservation and storage of biological material at very low temperature, usually at-196°C, the temperature of liquid nitrogen. At this temperature, the metabolic activities of the cells are arrested but they remain viable. Their normal functions can be reactivated after proper thawing. This method is useful for the long-term preservation of gametes. Following cryopreservation of Atlantic Herring (*Clupea harengus*) testes (Blaxter, 1953), the feasibility of successfully cryopreserving spermatozoa has been demonstrated in over 200 fish species (Billard *et al.*,1995a) notably for the Salmonids, Carps and Tilapias. Although a number of different protocols, even for the same species are advocated in the literature (Leung and Jamieson, 1991), the components of cryopreservation are the same.

Cryogenic agents such as solid CO_2 (-79°C) and liquid nitrogen (-196°C) are commonly used for freezing and storing of fish gametes. Nowadays liquid nitrogen is widely used for cryopreservation. Asynchronisation of gonadal maturation is one of the major constraints that impede fish breeding. To solve the asynchronisation of gonadal maturation, the preservation of gametes has emerged as a promising and a very useful technique to facilitate artificial breeding. Cryopreservation of fish spermatozoa is more preferred due to suitability of size for preservation, availability of large number of sperms per unit volume of milt (several millions per ml milt), ease of collection and handling. Sperm banks have been established for cultivated grouper, salmonids, Indian major carps and for endangered fishes. Unlike spermatozoa, attempts to cryopreserve fish eggs and embryos have met little or no success. The cryopreservation of fish ova and embryos is quite complicated when compared to fish spermatozoa. The reasons are **i)** large egg volume **ii)** the presence of two different egg membranes and the different water permeability of each of the membranes **iii)** sufficient degradation during cooling and **iv)** presence of large volume of yolk.

Advances in research on cryopreservation of gametes and embryos of aquatic organisms are modest compared with work done in terrestrial plants and animals. While sperm has been successfully cryopreserved in a number of cultured finfish and shellfish species, utilization at the farm level is very limited. Modest success has been achieved in the cryopreservation of shellfish embryos and early larvae. In the other hand, cryopreservation of finfish ova and embryos has not been successful so far. Unlike shellfish eggs and embryos, finfish ova and embryos are large, contain a large amount of yolk and are covered with a relatively thick chorion. Uniformity in the penetration of conventional cryoprotectants in cooling during the freezing process has not been attained in these large and dense specimens (Chao and Liao, 2001).

For any successful cryopreservation method, cells must survive freezing and thawing at such low temperature. Stability of cells for centuries requires storing them at temperatures below -130°C to ensure survival after the subsequent return to ambient temperature. Cryopreservation of sperm has been well established for many years in many finfish species but only in a limited number of shellfish. Most studies have been conducted on species that are of commercial importance. Knowledge of the biology of fish gametes is important for the development and design of cryopreservation procedures.

Gamete Preservation Techniques

In general, gamete preservation can be done in two ways:

1. Short term preservation (Non cryogenic)
2. Long term preservation (Cryopreservation)

Short term preservation is a pretreatment of experimental materials, followed by short term storage at 4°C, while cryopreservation involves short or long term storage of living cells or treated materials using liquid nitrogen as a coolant.

1. Short Term Preservation

The short term preservation is a technique in which the gametes are stored from a few hours to a few weeks in an unfrozen state. Short-term preservation of gametes can be done at temperature between 4 to 0°C. Washed spermatozoa sometimes possesses better capacity for short-term storage. The need to store gametes for short periods depends on several factors such as the country in which the hatchery is located, the size and type of aquaculture or conservational operation, the management of brood stock, methods of seed production and the principal purpose and markets for the seed. Similarly, if milt or ova of special interest are required for hybridization, they can be stored until fertilization is carried out.

A short-term cryopreservation of sperm of cyprinids (*Cyprinus carpio, Ctenopharyngodon idella* and *Hypophthalmichthys molitrix*) up to 12 days at 2°C to 6°C was possible without any remarkable reduction in quality when streptomycin was added (Jaehnichen, 1992). Chilled sperm of striped trumpeter (*Latris lineata*) were motile 8 days after collection (Ritarand Campet, 1995). All milt samples stored for 5 days on ice and replenished daily with oxygen retained at least 80 per cent motility and 99 per cent viability. In most fishes, however milt may only be stored for 2-3 weeks. After that, sperm motility remarkably decreases and currently, there is no method other than cryopreservation that can prolong the duration of sperm storage. A corresponding relationship between motility and fertilization capacity is observed in most fish species.

Ova stripped out soon after ovulation and stored in vitro in an undiluted state retain fertilizing ability for a few hours. The fertilizability of eggs can also be preserved by keeping them in the ovarian fluid or in an isotonic electrolytic solution for few hours to a few days at cold temperature above O°C (Harvey and Kelley 1984; Stoss and Donaldson, 1983).

2. Long Term Preservation

In long term preservation (cryopreservation), the gametes are stored along with suitable extender medium and cryoprotectants for longer duration in a frozen state at a very low temperature (-196°C). Spermatozoa can be stored for years together in liquid nitrogen (-196°c) where at this temperature all the biochemical activities of a living cell stops.

Crypreservation of Finfish and Shellfish Sperm

There are generally five major steps in cryopreservation procedures. The Steps can be modified depending on the species. The protocol for Cryopreservation of fish spermatozoa can be explained as follows:

Preparation of diluents (Extenders and cryoprotectants)

↓

Collection of fish spermatozoa

↓

Dilution of spermatozoa in extender and mixed with cryoprotectant

↓

Equilibration

↓

Ampouling

↓

Freezing

↓

Storing

↓

Thawing

a) Collection and Evaluation I

The key to successful preservation of fish gametes lies in the careful collection of gametes. Suitable techniques for milt collection reduce the spoilage of milt and external as well as internal injuries to the donor fish. The fish spermatozoa are collected from matured males either by stripping into a vial syringe or by squashing the testis. Especially, it is collected from species with asynchronous maturation habits. Slicing the testes within 10h postmortem to obtain a sperm suspension and inserting a micro capillary tube into the genital pore are two effective methods for fishes having scarce milt in addition to stripping (Chao *et al.*, 1992). Immediately after collection the milt is kept in an insulated box with ice and moved to laboratory. The age at maturation of experimental finfish varies with species. For example, it is 5 - 8 years for milkfish, *Chanos chanos;* 4 – 6 years for the Japanese eel, *Anguillajaponicus,* 2 - 3 years for grey mullet, *Mugil Cephalus* etc. For the shellfishes, a great maturation is 10 months for small abalone, *Haliotis diversicolor,* 6 - 10 months for pacific oyster, *Crassostrea gigas, etc.*Quality of milt changes depending on the pre monsoon and post monsoon period. Success of cryopreservation decreases in sperm collected toward the end of the spawning season. Therefore, the recommended time for sperm collection is species specific depending upon the location. In Taiwan, the best period to collect sperm for cryopreservation is from February to July for red belly tilapia, from December to February for black porgy, from December to January for grey mullet, and from April to July for malabaricus grouper (Chao, 1991). Regarding shellfishes, the period from September to November is suitable for small abalone, from January to September for pacific oyster and from May to August for hard clam (Chao *et al.*, 1997b). During the sperm collection, one has to pay attention to the following items in order to maintain sperm quality.

1. Collecting the sperm without contamination of feces, blood or scales
2. Holding the sperm during collection at the best condition by providing air or oxygen for respiration and
3. Maintaining the temperature of collected milt at 4°C during collection in the field and during transportation.

The quality can be evaluated by a series of physical observations such as volume of milt, spermatocrit value, sperm cell count and sperm motility. Sperm viability in terms of motility can be assessed using a microscope and a motility analyzer.The spermatozoa quality can be evaluated mainly by three parameters namely motility, concentration and fertilizing capacity. The motility analysis of spermatozoa is done by assessing the a) movement of the spermatozoa, b) duration of motility, c) percentage of motile spermatozoa and d) the measurement of the beat frequency of the flagellum.

The concentration of spermatozoa in the milt is measured by counting the number of spermatozoa in a haemocytometer or by spectrophotometric measurement. Spermatocrit value (volume of spermatozoa : volume of milt) is another approach. However, spermatozoa quality is ultimately tested by the fertilization and hatching rates.

Sperm fitness and in particular the inorganic and organic seminal plasma composition shows high intra and inter species variation (Rana,.1995 and 1995b). Semen quality analysis is based on milt samples expressed by abdominal pressure but unavoidable urine contamination can dilute milt by as much as 80 per cent and may result in false intra-individual variation. Such variability is likely to be further confounded by intra - male variation in protein and osmolality of urine (Rana, 1995a). The overall, resultant variability in milt may reduce sperm quality and alter cooling properties of the milt in an unpredictable and random manner.

Although it is widely recommended that spermatozoa be cryopreserved immediately after collection which may not always be practical under hatchery situations. Recent evidence supports the view that spermatozoa activated prior to cooling retain their potential to be successfully cryopreserved.

b) Equilibration with Extender and Cryoprotectant (Diluents)

i. Extender

An extender is a hypertonic solution of inorganic and organic chemicals, resembling the ionic composition of blood or seminal plasma, in which the viability of spermatozoa can be maintained prior to freezing. The chemical formulations of the extenders used for cryopreserving spermatozoa vary widely depending upon physiological and chemical characteristics of sperm. Extenders also possess a property that it initiates motility of spermatozoa when diluted with water. Wolf (1963) was the first person to develop an extender called Cortland medium, which is used for fish spermatozoa preservation. The other extenders which are commonly used include NaCI medium, KCI and NaCI + KCI medium. The proteins rich medium like egg albumin and Skimmed milk powder is widely used for the preservation of spermatozoa of fresh water fishes along with extenders. The functions of extenders are

1. To increase the volume of the milt/semen
2. To increase the viability of gametes *in vitro*

The chemical compositions of the extenders (Tables 17.1 and 17.2) used for cryopreservation is as follows:

Table 17.1: Chemical Composition of the Commonly Used Extenders

Sl.No.	*Extenders*	*NaCl (in µg/100ml of distilled water)*	*KCl (in µg/100 ml of distilled water)*
1.	NaCl medium	899	–
2.	KCl medium	–	1147
3.	NaCl + KCl medium	449.5	573.5

ii. Cryoprotectant

Cryoprotectant is a chemical, which protect the cell from chilling injury during the course of freezing. Blaxter (1953) was the first person to use glycerol as cryoprotectant. The optimum concentration of cryoprotectant varies from 5 - 15 per cent of the volume of diluents (extender + cryoprotectant) depending upon species.

Table 17.2: Chemical Composition of Extenders (Ingredients in mg/100 ml of distilled water)

Sl.No.	*Extenders*	*NaCl*	*CaCl H_2O*	*KCl*	*NaH_2PO_4. H_2O*	*$NaHCO_3$*	*$MgSO_4$. $7H_2O$*	*$Na_3C_5H5O_7$ $2H_2O$*	*Glucose*	*Fructose*	*Glycine*	*Na_2HPO_4*
1.	Cortland medium	725	23	38	41	1000	23	–	100	–	–	–
2.	Modified Cortland medium	188	23	720	41	100	23	–	100	–	–	–
3.	Alsevers Solution	40	–	–	–	–	–	80	205	–	–	–
4.	Fish Ringers Solution	650	–	20	30	20	–	–	–	–	–	–
5.	NaCl medium	899	–	–	–	–	–	–	–	–	–	–
6.	KCl medium	–	–	1147	–	–	–	–	–	–	–	–
7.	NaCl + KCl medium	449.5	–	573.5	–	–	–	–	–	–	–	–
8.	Glucose medium	–	–	–	–	–	–	–	5404.8	–	–	–

Cryoprotectants are added to a mixture of extender and milt to minimize the stress on cells during cooling and freezing. The cryoprotectants themselves inhibit the sperm motility. Although a number of cryoprotectants have been employed (Table 17.3), glycerol, dimethylsulfoxide (DMSO) and methanol are the most widely used cryoprotectants for preserving teleost spermatozoa. The widely used cryoprotectants and their best concentrations (Table 17.3) are as follows:

Table 17.3: Commonly Used Cryoprotectants and their Best Concentrations

Sl.No.	*Cryoprotectants*	*Best Concentration (per cent)*
1.	Glycerol	10-20
2.	Dimethyl sulfoxide	10-15
3.	Ethylene glycol	15-17.5
4.	Propylene glycol 10-20 v	10-20
5.	Methanol	15-12.5

There are two types of cryoprotectants namely,

a. *Permeating Cryoprotectants*

Permeating cryoprotectants are those which permeate into the cell membrane and give a protective lining. They help to reduce salt concentration during cooling and reduction in cell shrinkage. Glycerol, dimethyl sulphoxide (DMSO), ethylene glycol and methanol can permeate into the cell.

b. *Non-Permeating Cryoproctectants*

Non permeating cryoproctectants are those which depress the freezing point and raise the transformation temperature of the extra cellular solution. The non-permeating cryoprotectants like polyvinyl pyrrolidone (PVP), glucose, sucrose, egg yolk, serum and skimmed milk form a coating externally around the cell that prevents ice formation in its vicinity. The non-permeating cryoprotectant in conjugation with a permeating cryoprotectant, helps to depress the freezing point and prevents ice crystal formation in the vicinity of the cell. The adjuvant like egg yolk provides additional strength to membrane stability.

Diluent Ratio

The ratio of diluents (Extender+ Cryoprotectant) to spermatozoa varies from 1:1 to 1:20 and the mixing can be done at room temperature. The acceptable dilution ratio is, 1 part milt: 1 part extender for the sperm of grey mullet, black porgy and tilapia, 1:4 for milkfish and 1:20 for grouper. The cryoprotectant should not be mixed directly in the milt sample. It is mixed first with extender in the required ratio and kept in refrigerator. It is very much important to maintain the isothermal condition while mixing the diluents with milt. Equilibration is the time interval between mixing the milt in the diluents and putting the mixture into liquid nitrogen. It is mainly done to facilitate the penetration of cryprotectant into the cells. The longer the equilibration time the lower the fertility. Milt is usually diluted up to ten fold prior to cooling.

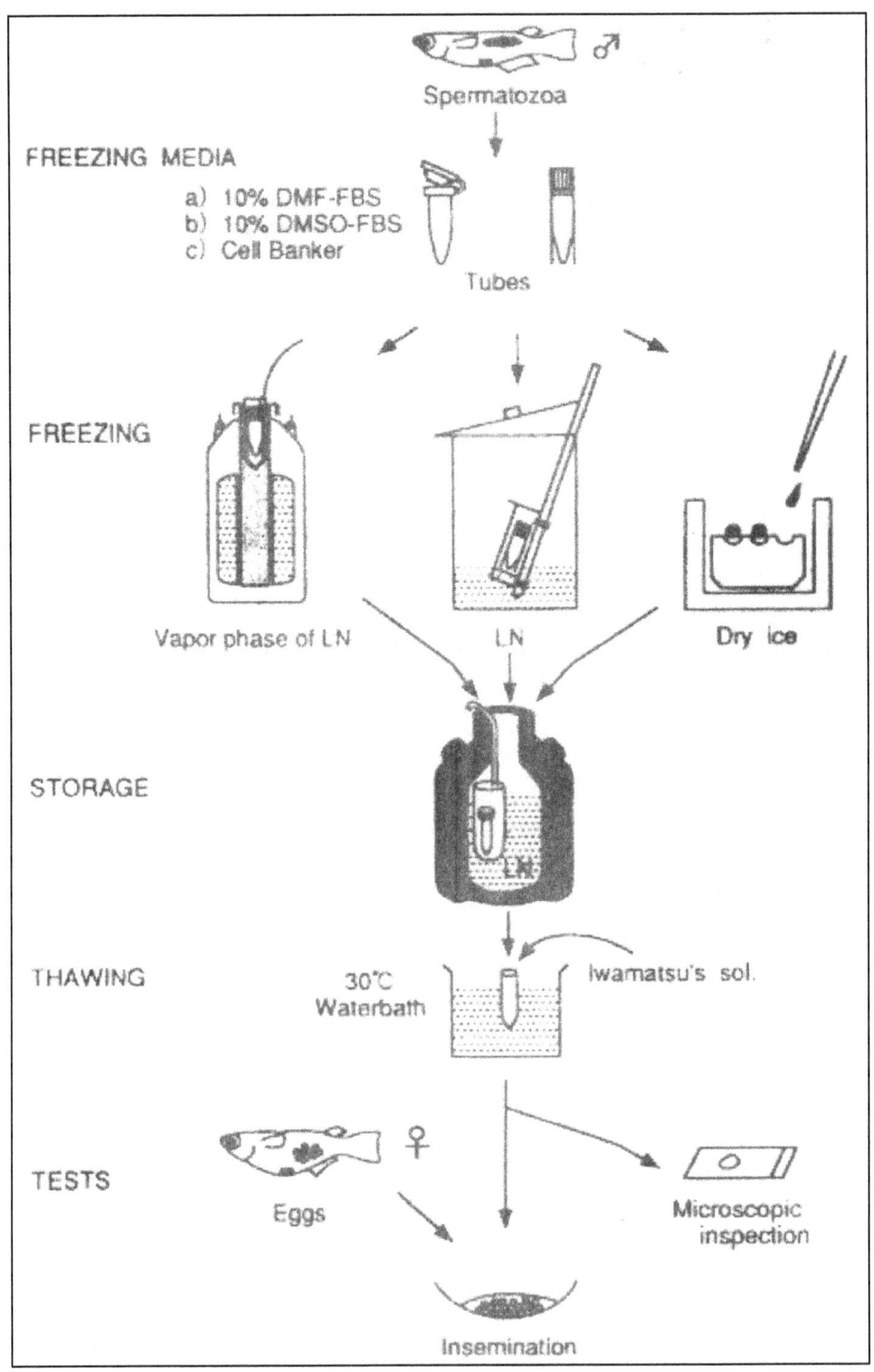

Figure 17.1: A Schematic Diagram of Cryopreservation Protocols of Sperm of a Finfish

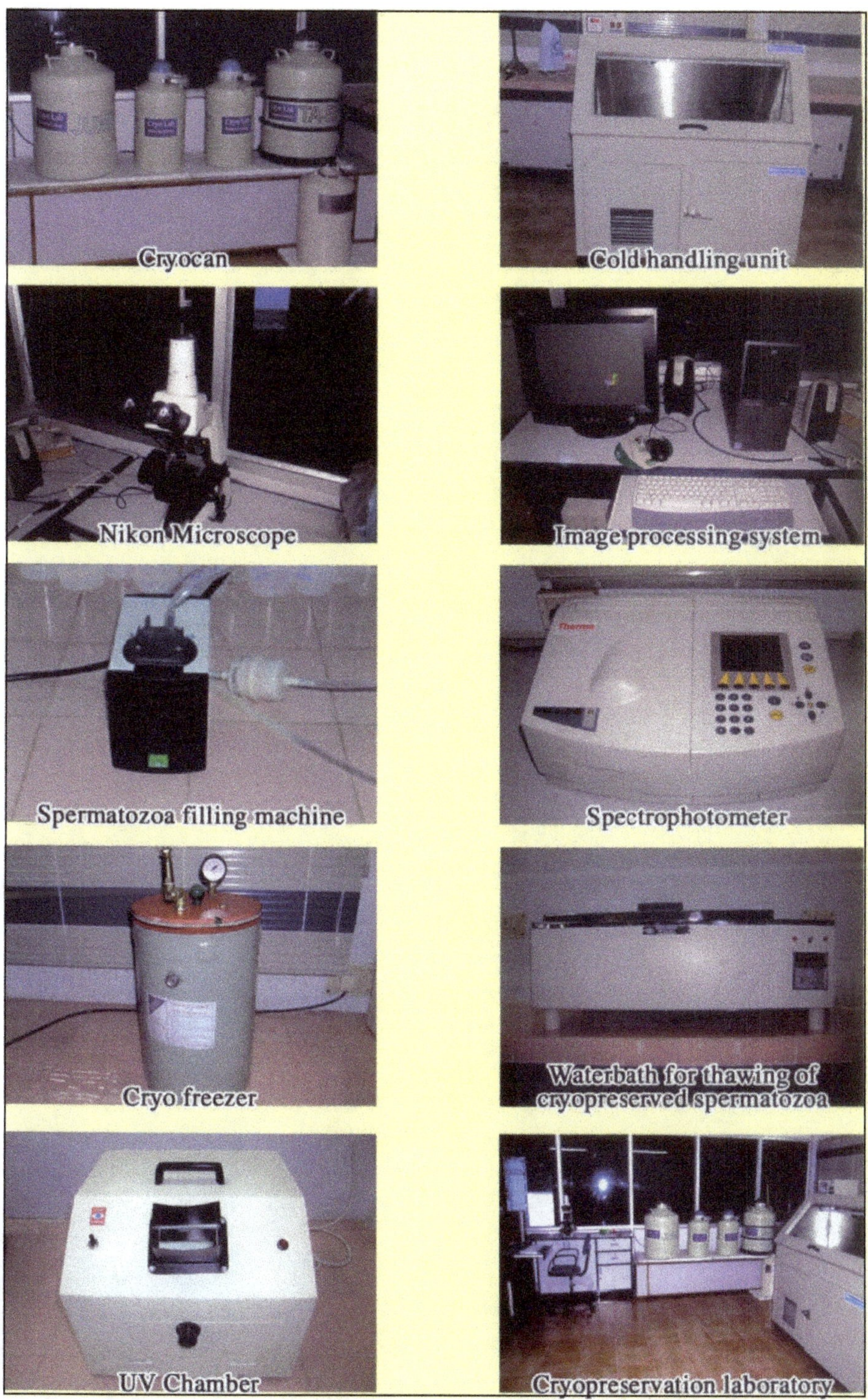

Figure 17.2: Lab Facility for Cryopreservation of Fish Spermatoza

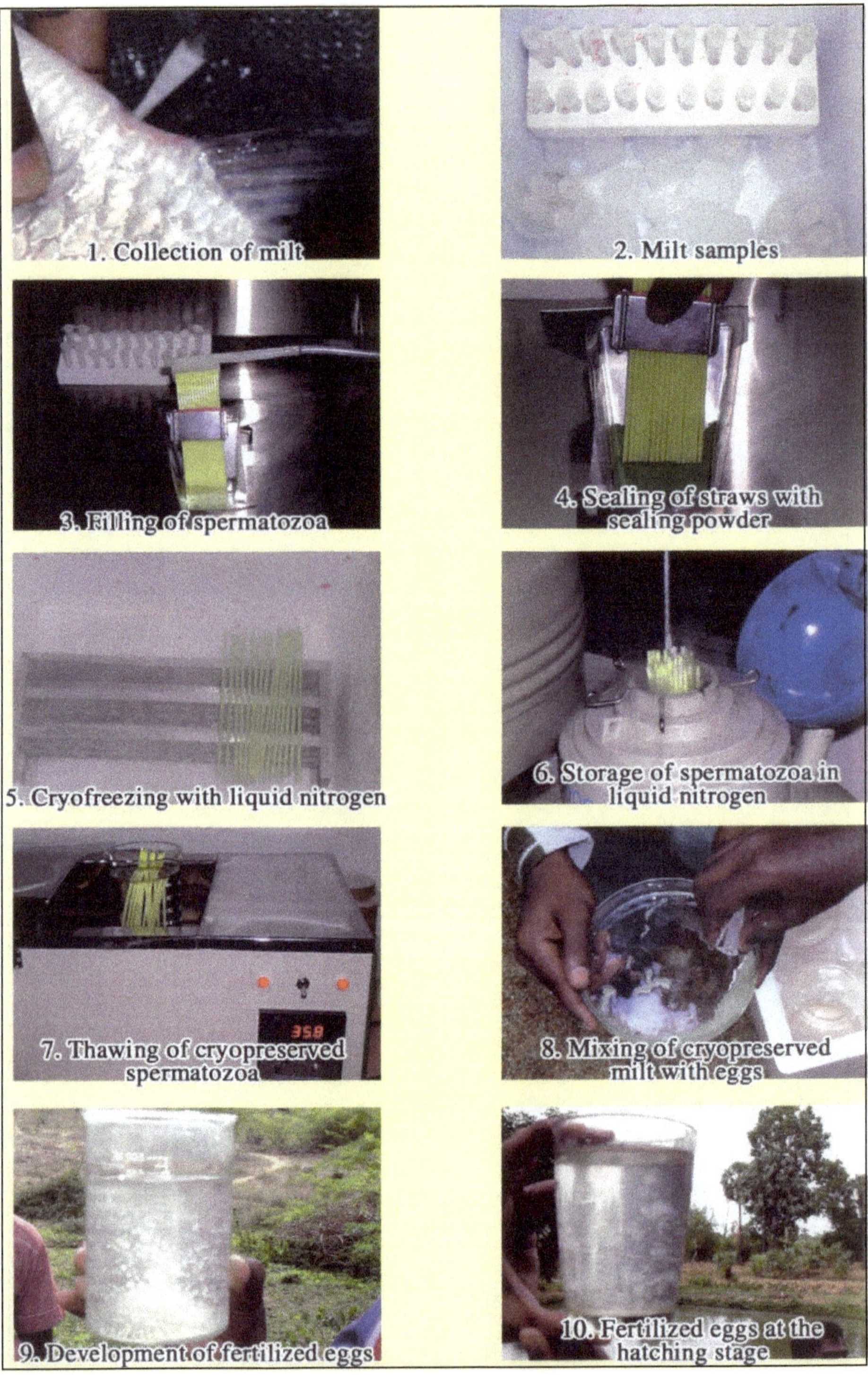

Figure 17.3: Cryopreservation Protocol

Sperm density, however, exhibits high seasonal and intra-individual differences (Leung and Jamieson, 1991). Consequently, sperm density can vary by as much as 300 per cent for any given dilution ratio. High cell densities may also result/in compression damage. During cooling water freezes and cells can be compressed in residual water channels. Therefore, in fish sperm cryopreservation standardized cell density should dictate the dilution ratio.

c) Freezing Protocol

The diluted samples within the equilibration period is filled in French straws and plugged with sealing powder. These are again stored under liquid nitrogen. Freezing protocol varies with species. The rate of freezing is a critical/factor during cryopreservation and each type of cell or tissue needs optimum freezing rate for the maximum post thaw survival rate.

For grey mullet, black porgy and grouper, two step freezing by exposing straws with sperm-cryoprotectant mixture to liquid nitrogen vapor at -100°C and then quickly quenching to -196°C is optimal (Chao *et al.*, 1992). For carp, the best freezing protocol is -5°C/min from 2°C to -7°C and then -25°C/min from-7 °C to -70°C, and for tilapias, -20°C/min from 25°C directly to -35°C and then-5 °C/min from -35°C to -75°C (Chao *et al.*, 1987). It appears that cooling rate in the/freezing phase has a very wide optimal range in the cryopreservation of sperm of finfish or shellfish. Different containers have varied cooling rates. The milt should be frozen immediately after collection. The freezing should be rapid so that thermal shock is minimal. This French straw filled with diluted sample is done by freezing manual vaporization over liquid nitrogen in thermocol chamber. Liquid nitrogen (-196°C) is the most commonly used cryogen for storing fish spermatozoa. The frozen milt straws are immersed in Liquid nitrogen and kept without disturbing. The use of the pellet technique for field cryopreservation, though convenient, is impractical for rational long term genetic resource banking. The shelf life of dry ice is short, particularly in the tropics, and perhaps more importantly frozen samples cannot be sealed to prevent possible cross contamination of disease in storage vessels. In addition, the thawing of large numbers of pellet is equally problematic. The fusing of pellets during thawing reduces the warming rates within aggregated pellets in an unpredictable manner (Rana, 1995).

During freezing several physico-chemical changes take place within the cell and its surrounding area. The plasma membrane that surrounds the cell gets affected by the cold shock. Plasma membrane consists of a lipid-protein bilayer and controls the transport of metabolites and ions. Lipid remains in a liquid state at normal temperature. At low temperature lipid gets frozen which affects the permeability and structural integrity of the membrane. Initially ice crystal formation occurs in the extra-cellular medium due to freezing. As a result, the extra-cellular medium becomes hypertonic to the cell. To maintain the osmotic balance, the intra-cellular water comes out leading to the reduction of the cellular volume. These changes cause mechanical damage to the cell. It is noted that the temperature range between 0 to -40°C is most critical during freezing since most of the cryo injuries take place in this temperature range.

Thawing and Fertilization

Cooling and thawing rates are regarded as the most critical phases of cryopreservation. Fast thawing is highly preferable than slow thawing since slow thawing may result in recrystallisation due to rehydration of cells. The thawing of stored sperm is accomplished by agitating them in water bath at warming rates between 50 and 70°C per minute. Preserved sperm is warmed directly from cryogenic temperature to the ambient water temperate at the time of fertilization. The water temperature during the spawning season is usually ideal for sperm revitalization. Cryopreserved sperm in Polyethylene straws can be plunged directly in water of sufficient volume at ambient temperature, which otters the optimal warming rate possible. Cryopreserved pellets are mixed in appropriate kind and volume of thawing solution at temperature higher than ambient temperature to reach an optimal final temperature. At the post-thaw phase, sperm that survived cryopreservation is ready for artificial fertilization. Cooling and thawing rates are the most important factors affecting the success of Cryopreservation. Fast thawing is more preferable than slow thawing because slow thawing can recrystallize the small intercellular crystals, which may damage the cells. Each french straws containing 0.5 ml thawed diluted milt is sufficient to fertilize 10,000 eggs. The thawed milt is mixed properly to the stripped eggs immediately and activated by a drop of water. The following cautions should be taken.

1. Determination of sperm motility as a guide to estimate their fertilization success
2. Shortening the time between thawing and fertilization
3. Dilution ratio

Thawing is required to bring back the cell to its normal temperature for resuming its usual function. During thawing the same physico-chemical changes occur as that of freezing but in a reverse order. Slow thawing produces large ice crystals which damage the cell structure leading to its death.

Resumption of Motility in Thawed Sperm

To improve motility and its stabilization in frozen thawed spermatozoa, experiments in common carp, *Cyprinus carpio,* were carried out using diluting solutions containing Na+ and K+ ions (Linhartand Cosson, 1997). A 275mM NaCl solution appeared to be best diluting solution which resulted in 80 per cent thawed spermatozoa motility against23 per cent spermatozoa motility of the control (undiluted) post-thawed sperm 10 min after storage. Apparently, solutions with pH close to neutral ensure motility in the cryopreserved sperm.

Fertility of Thawed Spermatozoa

Satisfactory fertilization rates have been obtained in thawed sperm in both finfish and shellfish. However, thawed spermatozoa may exhibit characteristics somewhat inferior to spermatozoa, which were not cryopreserved. Several studies have indicated that higher quantities of thawed milt are needed to obtain fertilization rates comparable to those when fresh milt is used. It is also suggested that the thawed spermatozoa could be activated using 1 - 2 per cent NaCl (Zhang *et al.*, 1994). Approximately 100 per cent motility could be restored upon thawing by adding a few drops of 5mM theophylline (Caylor *et a/.*,1994).

Applications

- ☆ Cryopreservation can provide a year round supply of seeds from desired species regardless of the spawning season.
- ☆ Cryopreservation helps in wider distribution of gametes in different hatcheries.
- ☆ It offers immense benefits to selection and cross breeding programmes/ and stock improvement.
- ☆ Disease free status of parent fish can be obtained through this technique.
- ☆ It reduces the need for a number of males to be maintained in the hatchery.
- ☆ It facilitates the conservation and propagation of endangered species in aquaculture practices.
- ☆ It helps in the management and conservation of aquatic resources
- ☆ The number of broodstock that has to be maintained in a seed farm can greatly be reduced.
- ☆ It helps in the study of androgenesis
- ☆ If the techniques for the cryopreservation of eggs/embryos/larvae are developed, then it can act as complementary to frozen sperm banks.

References

Billard, R., Cosson, J., Crim, L.W. and Suquet, M. 1995a. Sperm physiology and quality. Broodstock Management and Egg Larval Quality.N.R.Bromage and R.J.Roberts (Eds), Blackwell Science. pp: 25-52.

Blaxter, J. H. S. 1953. Sperm storage and cross-fertilization of spring and autumn spawning herring. *Nature* (London). 172: 1189-1190.

Caylor, R.E., Biesiot, P.M. and Franks, J.S. 1994. Culture of Cobia (Rachycentron canadum); cryopreservation of sperm and induced spawning, *Aquaculture*. 125 (1-2): 81 – 92.

Chao N.H. 1991. Fish sperm cryopreservation in Taiwan : technology advancement and extension efforts. *Bulletin Institut Zoology*. 16: 263-283.

Chao, N.-H. and Liao, I.C. 2001. Cryopreservation of finfish and shellfish gametes and embryos. *Aquaculture*. 197: 161–189.

Chao.N.H., Tsai. and Liao, I.C. 1992. Short and long-term cryopreservation of sperm and sperm suspension of the grouper, *Epinephelus malabaricus* (Bloch and Schneider), Asian Fish.Sci. 5: 103 -116.

Harvey B. and Kelley R. N. 1984. Short-term storage of *Sarorherodon mossambicus* ova. *J Aquaculture*. 37: 391-395.

Jaehnichen, H. 1992. Further improvement of artificial propagation of cyperinidae by short-term preservation of sperm. Proc. Sci. Conf. Fish Rep. '92. Res. Inst. of Fish Culture and Hydrobiol., Vodnany (Czech Rep.). pp: 92-97.

Leung, L.K.P. and Jamieson, B.J.M. 1991. Live preservation of fish gametes. Fish Evolution and Systematics: Evidence from Spermatozoa. B.G.M.Jamieson (Eds). Cambridge University, Press Cambridge. pp: 245 – 269.

Linhart, O. and Cosson, J. 1997. Cryopreservation of carp (*Cyprinus carpio* L.,) spermatozoa; the influence of external K^+ and Na^+ on post-thaw motility. Pol. Arch. Hydrobiol. 44 (1-2): 275 – 279.

Rana, K. 1995a. Preservation of gametes. In : *Broodstock management and egg and larval quality* (ed. by N.R. Bromage and R.J. Roberts), Cambridge University Press, Cambridge. pp: 53-76.

Rana, K. 1995b. Cryopreservation of aquatic gametes and embryos: recent advances and applications. Goetz, F.W., Thomas, P. (Eds.), Proceedings of the Fifth International Symposium, Reproductive Physiology of Fish. Austin, TX, USA, pp: 85-89.

Ritar, A. and Campet, M. 1995. Cryopreservation of sperm from striped trumpeter Latris lineate, Proc. Int. Symp. Rep. Physiol. of Fish, Austin, TX, USA. pp: 137.

Stoss J. and E.M. Donaldson. 1983. Studies on cryopreservation of eggs from rainbow trout (*Salmo gairdneri*) and coho salmon (Oncorhynchys kisutch). Aquaculture. 31: 51-65.

Zhang, L., Liu, X., Chen, S., Lu, D. and Guo, F. 1994. Effect of dimethylsulfoxide on osmotic pressure and survival rate of the sperm of several freshwater fish species. *Acta Hydrobiol. Sin.* 18 (4): 297–302.

Part IX

Fish Health and Diagnostic Biotechnology

Chapter 18

Molecular PCR Technique in Aquaculture Health Management

S. Felix

Fisheries Research and Extension Center,
Tamil Nadu Fisheries University,
Madhavaram, Chennai – 50, T.N.

Principle of Polymerase Chain Reaction (PCR)

Polymerase Chain Reaction (PCR) techniques was invented by Kary. B. Mullis in 1985 while working for Cetus Corporation, California for which he received the Nobel Prize in 1993. PCR is an in-vitro technique, by which a specific target sequence of a DNA molecule that lies between two regions of known sequence can be amplified by repeated enzymatic termocycling process. The impact of PCR technology was immediate and revolutionary and is now a well-established technique in the field of molecular biology.

DNA Molecule

Genes are carriers of information from one generation to the next and they are nothing but segments of DNA. In its native form, DNA exists as a double helix. This helix comparises of two strands of DNA running anti-parallel to each other, held together by hydrogen bonds. These hydrogen bonds are formed between complementary nitrogenous bases arranged sequentially on both the strands. The bases are Adenine (A), Guanine (G), Thymine (T) and Cytosine (C), Adenine always forms a double bond with thymine whereas guanine always form a triple bond with Cytosine. The bases are attached to a sugar molecule and each sugar molecule is connected to the adjacent sugar molecule by a phosphodiester bond. Thus, a unit of

DNA comprising of a phosphate group, a sugar and a base which is known as "nucleotide". The 5′ and 3′ carbon atoms are involved in the phosphodiester bond.

Therefore the sequence of arrangement of nucleotides on a single strand runs from the 5′ to 3′ direction. Thus for the formation of a double strand the two single strands must to complementary to each other and at the same time run in the opposite direction. That is, if one strand runs in the 3′ –5′ direction and if on one strand there is adenine or Guanine the other strand must have either Thymine or Cytosine respectively.

Primers in PCR

Primers are two short segments of DNA with the sequence complementary to the regions flanking the target. In order the amplify a specific region of the genome it is not necessary to know the entire sequence of the target. Only the sequence of the regions flanking the target is necessary. The primers are then synthesized based on these sequences. The purpose of the primer is to provide a free 3′OH end form where the strand synthesis begins.

Major Steps Involved in PCR

There are three major steps involved in PCR *viz.* Denaturation, Annealing and Extension. The processes taking place in each of the three steps are described below.

1. Denaturation

This is the first step where the double stranded template DNA is subjected to a temperature of 94°C, as a result the hydrogen bonds between the two stands are broken and the DNA becomes single stranded. This step is usually done for 30 – 60 seconds.

2. Primers Annealing

In this process, the temperature is lowered to around 55°C. At this point the primers anneal to the template at a region where the sequence of the template is complementary to that of the primer. This process is known as 'Priming'.

3. Extension

In this step the temperature is increased to 72° C during which the themostable DNA polymerase enzyme will catalyze the catalyze the formation of anew strand by adding fresh nucleotides, starting from the 3′ OH end of the primers. Thus a new complementary strand is synthesized for each strand.

These three steps *viz.* Denaturation, Annealing and Extension constitute one cycle. Each of these step lasts for about 30secs to I minute which may some times depend on the target sequence to be amplified.

Production of Multiple Copies of DNA Fragments

Theoretically if the PCR cycle started with one template, at the end of one cycle there will be a total of two sets of similar target DNA molecules. For further cycles these double stranded DNA molecules will serve as templates. This process of amplification takes place as explained below:

When the extension product of each primer is sufficiently long, it will include the sequence complementary to the other primer. At 94°C the newly synthesized strand detach from the template and at annealing temperature more primers will hybridize at their respective positions, including the positions on the newly synthesized strands. The DNA polymerase enzyme now carries out a second round of DNA synthesis. The cycle of denaturation, primer hybridization and synthesis is repeated 25-30 times resulting in the eventual synthesis of several hundred million copies of the amplified DNA fragment.

The extension products themselves act as template in future cycles, resulting in the exponential increase in PCR product with the number of cycles. The first extension products result from DNA synthesis on the original template do not have a distinct length as the DNA polymerase will continue to synthesize new DNA until it either stops or is interrupted by the start of the net cycle. In the second cycle also the extension products are of intermediate length. However, at the third cycle fragments of target sequence are synthesized which are of defined length corresponding to the position of the primers on the original template. From the 4th cycle onwards, the target sequence is amplified exponentially. The amplification, as a final number of copies of the target sequences can be expressed by the following relationship:

$$\text{Final number of copies} = (2n - 2n)\ N$$

where,

N: Number of copies of the original template.

n: Number of cycles

2n: First and second cycle products of undefined length

If the PCR run is 100 per cent efficient during each cycle, after 20 cycles, there will be 220fold amplification. However, in practice only 20-30 per cent efficiency is attained auctioned in PCR runs. However, this will not normally affect the results. The concentration of constituents and primers in the PCR mix and the PCR conditions must be optimized for each test. The type and amount of template in the PCR mix must also be worked out accurately to get precise results through PCR.

Components of PCR

To amplify a desired target sequence using Polymerase Chain Reaction (PCR) technique the following five important components are required.

1. Primers
2. Buffers
3. Thermostable DNA polymerase enzyme.
4. Deoxyribonucleotide triphosphates (dNTPs)
5. Template DNA

1. Primers

A primer consists of a short single stranded oligonucleotide, which hybridizes to complementary sequences on the template flanking the target region. The primer has a free 3'OH end to which the DNA polymerase enzyme add nucleotides and thus serves as a starting point for the synthesis of a new strand. The optimum length of a primer is suggested as 18 – 25 bases. Too short primers may hybridize to non-target sequences and give undesired amplification products. Primers of longer length, hybridizes with template DNA at a much slower rate, thereby reducing the efficiency of the process. The primers in general should not be complementary to each other. If so, they will form an artifactorial product called "Primer-Dimers". In designing primers, unusual sequences such as stretches of polypurines or polypyrimidines should be avoided. Primer should be of similar length and have a matched GC content between 40-70 per cent. Primers are usually used at a concentration of 1μM, which is sufficient for at-least 30 cycles of amplification. The presence of higher concentration of oligonucleotides can cause priming at ectopic sites with consequent amplification of undesirable non-target sequences. Conversely, the PCR is extremely inefficient if the concentration of primers is limiting.

2. PCR Buffers

One factor that influences the efficiency of PCR is the buffer composition. Several formulations of PCR buffer is available of which the commonly used composition is as follows. The concentrations specified are for a buffer having strength of 10X.

Component	*Concentration*
Tris-HCl	100 mM, pH 8.5 at room temperature
KCl	500 mM.
$MgCl_2$	15 mM
Gelatin/Bovine Serum Albumin	0.1 per cent (W/V)

i. Tris HCl

The most commonly used buffer for PCR is Tris HCl. For individual reactions the concentration of Tris to be used is 10mM. This buffer has a pH range of 8.5 to 9.0 at 25?C. The pH of the Tris buffer decreases at the rate of 0.3 units for each 10?C rise in temperature. Thus when the temperature increases to 72?C during extension in the PCR cycle, the pH will reach a value between 7.3 to 7.5 which is found to be the optimum pH for the DNA polymerase enzyme activity.

ii. Mg^{2+} ions

The Mg^{2+} ions function as cofactors for the DNA polymerase enzyme. Apart from that Mg^{2+} forms soluble complexes with the dNTP's which is found to be essential for dNTP incorporation. Mg^{2+} is therefore supplied as magnesium chloride in the buffer.

The concentration of the divalent cation Mg^{2+} is a critical factor. A very low concentration of Mg^{2+} reduces the activity of the enzyme. This is because of the dNTPs, which bind the Mg^{2+} ions. On the other hand excess Mg^{2+} concentrations is found to yield nonspecific products. Therefore the concentration of Mg^{2+} should be optimized whenever a new combination of target and primer is first used or whenever concentration of dNTPs or primers is altered. Optimum Mg^{2+} concentration in the form of Magnesium chloride is in the order of 1.0 – 1.5mM.

It is important that the reaction mixtures does not contain high concentration of chelating agent such as EDTA which will result in the chelation of Mg^{2+} or of negatively charged joint group such as phosphates.

iii. Potassium Chloride (KCl)

KCl is added to increase the efficiency of Primer annealing. The allowed concentration of KCl is 50mM. An excess concentration of KCl is found to inhibit the enzyme's activity.

iv. Gelatin or Bovine Serum Albumin

It is added as enhancers to the buffer. These agents are found to prevent the precipitation of the Taq DNA polymerase enzyme.

3. DNA Polymerase

DNA polymerase is an enzyme that catalyzes the extension of a polynucleotide chain using one of the original parental strands as template. This enzyme can only extend a polynucleotide chain and cannot synthesis an entirely new strand. Thus the enzyme requires a start point which should necessarily be a 3′ OH end of a DNA molecule. Therefore a primer is required for the enzyme to begin the synthesis. The enzyme functions by adding deoxyribonucleotides to the free 3′OH end of the primer. Therefore the chain grows in the 5′ – 3′ direction. Successive addition of the dNTPs to the primer results in the formation of a strand complementary to the template strand.

Thermostable enzymes are mostly preferred for PCR because thermolabile enzymes get inactivated at 94oC during the denaturation step. *Taq* DNA polymerase is a thermostable enzyme produced by the bacterium *Thermus aquaticus* found in hot springs. This enzyme can tolerate temperatures upto 110°C and is therefore widely used for PCR. Taq polymerase functions optimally at 72°C over a pH range of 7.0 – 7.5. The enzyme adds nucleotides at the rate of approximately 100 nucleotides per second. About 0.5 to 2.5 units of the enzyme is defined as the optimum range for a PCR reaction.

4. Deoxyribonucleotide Triphosphates

The basic units of the DNA molecule are supplied in the form of deoxyribonucleotide triphosphates (dNTPs). The four types of nucleotides used normally are;

- ☆ deoxy Adenosine triphosphate (dATP)
- ☆ deoxy Guanosine triphosphate (dGTP)

- ☆ deoxy Thymidine triphosphate (dTTP)
- ☆ deoxy Cytidine triphosphate (dCTP)

These dNTPs are incorporated in the growing DNA stand complementary to template DNA. Usually a stock solution of dNTPs at a concentration of 50mM is prepared and its pH is adjusted to 7.0 using 1N NaOH. This will ensure that the pH of the final reaction does not fall below 7.1. Nucleotides tend to disintegrate at high pH and lose its property when exposed to light for longer times. Therefore it is advisable to store the dNTPs at –20oC in the dark. DNTPs of high purity are commercially available.

5. Template DNA

The template DNA may be prepared from a variety of tissues and cell types. The samples may be single or double stranded DNA for the PCR and RNA in case of "Reverse Transcriptase PCR" (RT-PCR). For diagnosis of WSSV in shrimps, pleopods, gills, hemolymph and pereiopods can be used as a source for obtaining the template. Eyeballs and hepatopancreas should not be used as they have been found to give inconsistent results, which might be due to the presence of 'PCR inhibitors'. The hepatopancreas, being a digestive organ, contains DNase, which can digest the DNA.

The concentration of template DNA is a critical factor in setting up the PCR. Too many templates will inhibit PCR by reducing the amplification efficiency and also causes smear problem. The template concentration between 200 - 1000ng is found to be optimum, contrarily fewer templates can lead to a drop in the sensitivity.

Equipment Required for a PCR Laboratory

Apart from the Standard laboratory accessories the following equipment are indispensable for a commercial PCR laboratory.

1. Thermal Cycler/DNA engine.
2. Microcentrifuge
3. Agarose Gel Electrophoresis apparatus
4. U.V. Transilluminator
5. Polaroid Gel Camera (or) gel documentation system

1. Thermal Cycler

The thermal cycler is an instrument, which is used for the amplification of the template. A unique feature of this equipment is that it can be programmed to provide varying temperatures for a specified duration. The thermal cycling equipment has been developed based on the principle of PCR. Several models of equipment with advanced features exist. A normal thermocyler will have the following facilities.

- ☆ Sample block to accommodate different types of PCR tubes, microplates and slides (for *in-situ* PCR).
- ☆ Programmable software to suit the requirements of each cycle. It also includes facilities for editing, file management, password protection, etc for various thermal cycling parameters.

- ☆ Most Thermal cycling equipment have in built programmes for certain standard protocols such as for incubation, Denaturation, etc.
- ☆ Hot bonnet heated lid for oil-free cycling
- ☆ Space saving design for easy setup and transportation

To ensure consistent results the thermal cycler must be operated as suggested in the manual provided by the manufacturer. The thermal cycler should be placed in a spacious area, which must be air-conditioned. Since the process involves high temperatures care must be taken to avoid serious incidents.

2. Microcentrifuge

The microcentrifuge is used at various stages during the nucleic acid extraction protocol. Two types of the centrifuge will be necessary. A refrigerated centrifuge with an rpm range of 5000-20,000 rpm is required to keep the samples at low temperature (about 4°C) during centrifugation. Normally in non - refrigerated centrifuges the temperature of the unit is found to increase during longer duration runs. This increase in temperature may cause the denaturation of the sample.

The other one will be a tabletop centrifuge having an rpm range of 1000 to 10000 rpm. This centrifuge may be required for brief spinning of samples and therefore need not be refrigerated. Care should be taken while using a centrifuge which otherwise may affect the efficiency and shelf life of the instrument.

Micropipettes and PCR Tubes

The hand of micro quantities of reaction components requires the use of sterlizable micropipettes and PCR tubes. It would be advisable to maintain a set of micropipettes covering a range of 0.5µl to 1000µl. separate pipettes should be maintained for addition of PCR reaction components. Disposable pipettes tips must be used as far as possible. Positive - displacement micropipettes with disposable tip plungers are highly effective but expensive. It contamination becomes a problem, an added level of security can be ensured by the use of tips, which are supplied fitted with cotton, this will help to prevent cross contamination.

There are many varieties of PCR tubes. Choice is based on convenience of use, availability, cost type of machine in use, etc. a number of suppliers have developed PCR tubes having special properties such as thin walled typed, once with flat lid etc. Basically the tubes should be easy to use and fit well into the heating block.

3. Agarose Gel Electrophoresis Apparatus

Electrophoresis is used for the identification and determination of molecular weight of proteins and nucleic acids. Therefore electrophoresis equipment should be available in a commercial PCR laboratory. This can be used to visualize the PCR products. Generally a submarine gel electrophoresis model is used for nucleic acid analysis. A suitable and compact electrophoresis model should be chosen as per ones requirements. UV transparent model allows for the visualization of the nucleic acids without requiring to handle the gel. This can help prevent contamination by the ethidium bromide stain.

The electrophoresis process requires a suitable powerpack, which can provide undisturbed supply of current at a constant voltage. Standard pwerpacks allow for two or more simultaneous runs. This will be useful especially when large number of samples have to be analyzed.

4. Ultra Violet (U.V.) Transilluminator

UV Transilluminator is used to visualize the DNA/RNA fragments in gels. Most transilluminator provide high quality electronic single intensity or variable intensity UV light. The Transilluminator consist of lamps emitting UV light of wavelengths 300 nm or 365 nm. The emitted UV light is evenly distributed at the filter surface for better identification and documentation of viewed gels. The 300nm UV induced high fluorescence and sensitivity in ethidium bromide/DNA complexes producing sharp, distinct bands. At this wavelength, photonicking, dimerization and photo bleaching of gels are minimized. The UV intensity at the filter surface is between 8000 to 9000 microwatts/cm^2. the UV light is dangerous to the eyes and skin and therefore extreme caution must be exercised while using the equipment. UV resistant gear such as gloves and goggles must be worm to avoid damage by UV light. The arrangement on the transilluminator is such that a Polaroid gel camera can be mounted and used for obtaining pictures of the gel.

5. Polaroid Gel Camera

The gel camera is designed to take instant pictures of the gel. The camera can be mounted permanently on to the transilluminator or hand held. Polaroid negatives are required for the purpose. Photos of the results may not be required in a commercial set up but in rare cases when required it can be used. The working of the camera is just as the ordinary camera excepting that it has a special filter to facilities fluorescent light emitted by the nucleic bands.

6. Gel Documentation System

The gel documentation system is a compact equipment to analyze the gels. It contains a software programme, which can be used to modify the view of DNA/RNA bands in gels. The advantage of using the gel documentation system is that there is no danger of exposure to the UV light. The equipment analyses the gel by itself and provides photographs. There is also a monitor, which shows a direct pictures of the gel. The intensity of the gel can be adjusted to provide good quality pictures. The information can be stores in suitable retrieval system *i.e.* floppy discs etc or can be alternatively stored in the system itself.

The equipment is very costly and can be substituted by an ordinary UV transilluminator for commercial purposes.

Other Equipment

The laboratory will require certain general equipment along with the above mentioned ones, such as:

Deep freezer (-20°C/-80°C), refrigerator, Hot waterbath, Vortex mixer, Gel rocker, Triple glass distillations unit, Magnetic stirrer, pH meter, Hot air oven, Autoclave,

Microbalance, Laminar airflow cabinet, Manual tissue homogenizer, Hot plate, Micro oven, etc.

Agarose Gel Electrophoresis

Electrophoresis is a molecular technique by which charged molecules in solution, chiefly proteins and nucleic acids are separated as a result of the difference in their individual molecular weight during migration in response to an electric field. Agarose gel electrophoresis is performed in agarose gel matrix so that molecules of similar electric charge can be separated on the basis of size. Their rate of migration or mobility through the electric field depends on the strength of electrical field, the net charge, size and shape of the molecules, the iconic strength, temperature and viscosity of the medium in which the molecules are moving. Proteins are amphoteric compounds, that is they contain both acidic and basis residues. Their net charge is determined by the pH of the medium they are in. Unlike proteins, nucleic acids are not amphoteric. They remain negatively charged at any pH and hence emigrate towards anode during electrophoresis. The heavier DNA molecules (more weight) move position in an electrical filed, these molecules will resolve into bands which comprises molecules with same charge to mass ratio.

Materials and Equipment

1. Agarose
2. Electrophoresis apparatus and Powerpack.
3. Tris base
4. Boric acid/Acetic acid
5. Gel loading buffer
6. DNA Marker
7. Ethidium Bromide stain (5mg/ml)
8. EDTA
9. Triple glass distilled water
10. Sample
11. Micropipette

Agarose

Agarose is a highly purified polysaccharide derived form agar. Unlike agar, agarose is not contaminated with charged material. Agarose, which is supplied in powder form, dissolves when added to boiling liquid. It remains in liquid state until the temperature is lowered to about 40°C at which it gels. The gel is stable; it will not dissolve again until the temperature is raised back to about 100°C. the pore size of the gel can be predetermined by adjusting the concentration of agarose. Higher the concentration, the smaller the pore size and vice versa. The working concentrations are frequently in the range of 0.4 to 2 per cent w/v. Agarose gels are used for analyzing a wide range of size variations. The separation characteristics of agarose gel at various concentration are given below:

Percentage of Agarose	*Efficient Range of Separation (kb)*
0.3	60 - 05
0.6	20 - 01
0.7	10 - 08
0.8	10 - 05
0.9	07 - 0.5
1.2	06 - 0.4
1.3	04 - 0.2
2.0	03 - 0.1
3.0	00 - 70

Agarose gels are fragile; they are actually hydrocolloids and are held together by the formation of week hydrogen and hydrophobic bonds. For this reason, an agarose gel should be handled with special care. Careless handling of them can cause the gel to tear.

Buffers

Agarose gels are usually prepared using Tris- Acetate – EDTA (TAE) or Tris – borate – EDTA (TBE) buffer. These buffers are prepared as follows:

TAE Buffer – Stock Solution (50x)	*TBE Buffer – Stocked Solution (10x)*
Tris base – 242 gm	Tris base – 108 gm
Glacial acetic acid – 57.1ml	Boric acid – 55 gm
0.5m EDTA – 100ml	Disodium EDTA – 9.3 gm
pH – 8.0	pH – 8.2
Triple distilled water – 1lit.	Triple distilled water – 1lit.
Working Solution (1x)	*Working Solution (1x)*
TAE buffer (50x) : 20ml	TBE buffer : 100 ml
Triple distilled water : 980 ml	Triple distilled water 900 ml

Preparation of the Gel

Appropriate quantity of agarose depending on the percentage of the gel required is mixed with 1 x buffer and boiled unit all agarose is dissolved. After cooling it to 55°C, the contents are poured into the gel tank already fitted with the combs. On cooling the agarose solidifies. The combs are carefully removed to expose the wells. The gel tank is then filled with 1x TAE or TBE buffer. The gel is now ready for loading the sample.

Gel Loading Buffers

Before loading DNA samples, they are mixed with gel loading buffer. The most commonly used gel-loading consist of 0.25 per cent bromophenol blue, 0.25 per cent

Xylene cyanol are used to identify the migration of sample molecules and sucrose is used to increase the sample density, thereby facilitating effective sinking of samples into the wells. Instead if sucrose, 30 per cent glycerol or 15 per cent ficoll can also be used.

Loading the Samples

The reservoir tanks of the electrophoresis apparatus must be filled with 1x buffer. Using disposable micropipette and gel loading buffer different samples are introduced in wells formed in gels. Usually 3L of 1x gel loading buffer is mixed with 5?l of the sample on a parafilm and then the sample is loaded into the wells without spilling. 5l off a suitable marker DNA mixed with 3l of gel loading buffer in loaded in the adjacent well and run simultaneously so as to determine the sizes of respective bands. A marker consists of DNA fragments of know sizes. After loading the samples, the electrophoresis apparatus is connected to power supply. Usually, the gel is run at 5 – volts/cm of the gel, which is sufficient fir effective separation of the DNA bands.

Analysis of the Gel

To view the DNA bands the gel has to be stained using Ethidium Bromide. Staining with ethidium bromide can be done in two ways.

i. Adding to the Gel Directly Before Sasting

Before pouring the gel into the tray 2-3l of 5mg/ml Etbr is added to the gel and shaken well. After sufficient cooling the gel is then poured in to the gel tray. This is the easiest and quickest methods. But has the disadvantage that the buffer gets contaminated and ethidium bromide is a serious carcinogen.

ii. Alternatively the Gel can be Run Without Adding Etbr to it

After the run the gel is taken separately and put in a gel – staining tray. To the tray containing the gel 100 ml of distilled water is added. To the distilled water 5l of 5mg/ml Etbr is added. The gel is then shaken on a gel shaker. After 30 min the distilled water is discarded and fresh distilled water is added. The gel is shaken in the fresh distilled water for destaining. The gel is then taken out and viewed. This method is time consuming, but the contamination of the buffer and gel tray by Etbr can be avoided.

Visualization of DNA

At the end of electrophoresis, the gels could be visualized under UV Transilluminator. The fluorescent dye, ethidium bromide in the gel binds to DNA by UV light. Thus DNA in the gel are visualized. About 5ng of DNA can be detected with ethidium bromide. Since ethidium bromide is a potent carcinogen, it must be handled cautiously. Comparing the molecular weight of the band of the marker the molecular weight of the band(s) of the sample can be assessed.

Template DNA Preparation for PCR

Extraction of nucleic acid from the sample for the PCR is the most important step involved in PCR protocol. Inadequate attention in this step could lead to false negative results. However PCR technique can to some extend, tolerate poor quality templates. Hence isolation and purification of the viral nucleic acid using density gradient and ultracentrifugation techniques is not essential.

To obtained the template for PCR analysis conventional methodologies to isolate genomic DNA is found to suffice, since one can expect the viral DNA (*i.e.* the template) to be along with the genomic DNA of the infected host. This nucleic acid is a unique feature of the PCR technique.

Several protocols exist for the extraction of template DNA. The conventionally used method involving Proteinase K digestion followed by Phenol- Chloroform extraction is found to yield good quality templates, but has the disadvantage that it requires a considerable amount of time (about 1-2 days). On a commercial basis, immediacy of results is crucial and hence methods aimed at overcoming such factors have to be adopted. The rapid boiling method and the use of CHROMA SPIN + TE = 1000 columns are two such methods for rapid results. The former method though rapid, is found to yield poor quality templates. Whereas the latter combines both efficiency and yield and may be a suitable method for rapid template extraction on a commercial basis. All the three methods have been outlined below. The choice of the method rests on the suitability of its application. Sources of template.

For the purpose of DNA isolation the pereiopods, gills, tissue covering the inner surface of the carapace over the gills, the eyestalk (without the eye), etc. can be used a source for template DNA extraction. In the case of fish, tissues can be used. Hemolymph has also been suggested as good source for viral template DNA. Based on the type of source used one or more of the methods can be used.

Materials and Equipment

- ☆ Shrimp samples: adult, juveniles, post larvae, hemolymph etc.
- ☆ Micropipettes (ranges): 0.5 - 2µl, 2-20µl, 20-200µl and 100 – 1000 µl.
- ☆ Refrigerated centrifuge – (20000 x g)
- ☆ Table top centrifuge – 10000 rpm
- ☆ Serological waterbath
- ☆ Tissue homogenizer
- ☆ Microfuge tubes (1.5ml)
- ☆ Saturated Phenol
- ☆ CHROMA SPIN + TE = 1000Columns
- ☆ Disposable syringes
- ☆ Vortex mixer

Reagents and Chemicals

- ☆ Absolute ethanol
- ☆ Chloroform
- ☆ Double Glass Distilled Water (DGDW)
- ☆ Ethylene diamine tetra acetic acid (EDTA)

- ☆ Isoamyl alcohol
- ☆ Proteinase K
- ☆ Sodium acetate
- ☆ Sodium Dodecyl Sulfate
- ☆ Sodium chloride
- ☆ Sodium hydroxide
- ☆ Tris Hydroxyl Methyl Amine – Tris

Note: All reagents and chemicals are to be molecular biological grade

Protocol 1

Template DNA extraction using Proteinase K digestion and Phenol Chloroform Method (After Otta *et al.*, 1999)

1. Chop the pereiopods/or appropriate tissue into fine pieces using a sterile surgery blade in a sterile petridish.
2. Transfer the finely minced pieces of tissue (approximately 400mg) into 1.5ml Microfuge tubes.
3. Using a tissue homogenizer, homogenize the tissue into fine particles in the Microfuge tubes.
4. Spin the Microfuge tubes in a tabletop microcentrifuge at 2000 rpm briefly to pool the fine macerated particles sticking on the walls of the tubes.
5. To each of the homogenized tissue add 500µl of Lysis buffer (containing 100 mM NaCI, 10mM Tris HCI pH 8.0,25 mM EDTA ph8.0, 0.5 per cent SDS, and 0.1mg/ml of Proteinase K).
6. Incubate the Microfuge tubes in a waterbath at 55°C overnight (8-10h)
7. After incubation, re-homogenize the samples using a tissue homogenizer and then centrifuge the samples at 10000 rpm for 10 minutes.
8. Transfer the supernatant to fresh Microfuge tubes using a micropipette.
9. To the supernatant, add an equal volume of Phenol Chloroform and Isoamyl alcohol in the ratio of 25:24:1.
10. Gently mix the contents by inverting the Microfuge tubes for 5 minutes and then centrifuge at 5000 rpm for 5 minutes.
11. After centrifugation, transfer the supernatant (aqueous phase) using a micropipette into another microfuge tube without disturbing the aqueous and the organic phases.
12. Repeat Phenol: Chloroform: Isoamyl alcohol extraction twice.
13. To the supernatant thus obtained, add equal volume of Chloroform to remove additional contaminating proteins and traces of phenol. Gently invert the microfuge tubes to mix the contents. Then spin the centrifuge tubes at 10,000rpm in a tabletop centrifuge at room temperature for 10 minutes. Repeat this step thrice.

14. Transfer the supernatant obtained to a fresh 1.5ml centrifuge tube. Then add 1/10th volume of 3M sodium acetate and 3 volumes of chilled absolute ethanol. Store at -20°C overnight to precipitate the DNA.
15. Spin the microfuge tubes in a refrigerated centrifuge at 17000xg for 30 min at 4°C to pellet the DNA.
16. After centrifugation discard ethanol and wash the pellet with 500µl of 70 per cent ethanol and again centrifuge at 17000xg for 15 minutes at 4°C. Repeat this step once more.
17. Discard ethanol and allow the DNA pellet to dry.
18. To the dried DNA pellet add 100 µl of sterile DGDW and place at 37°C for 12h to dissolve the DNA.
19. 1-2 µl of the dissolved DNA can be used for template for PCR analysis.

Protocol 2

Template DNA Extraction using CHROMA SPIN + TE – 1000 (Clontecha) Column (TVMDL, Texas A&M university, USA):

A. Extraction of Template DNA from Hemolymph

1. Take 2 drops of Hemolymph using a disposable syringe and transfer it into 1.5ml microfuge tube.
2. Add 350 ml of Lysis buffer (50mMTirs, 20mM EDTA, 0.5 per cent SDS pH 8.0) and 20ml of Proteinase K (20mg/ml) to the sample.
3. Incubate the sample at 65°C for one hour and boil the sample in a boiling water for 10min.
4. Centrifuge the sample at 16000xg for 3min.
5. Place the end of the CHROMA SPIN + TE – 1000 (Clontecha) Column into a 1.5ml microfuge tube.
6. Take 100ml of the supernatant from strep 4 and transfer it slowly in to the CHROMA SPIN – 1000 Column.
7. Centrifuge the column in a tabletop centrifuge at room temperature for 5 min at 1500rpm.
8. Discard the spin column and the collect the purified sample for the bottom of the microfuge tube.
9. The purified sample is suitably diluted (1 or 2x 10) for use in PCR analysis.

Protocol 3

Template DNA isolation from hemolymph

1. Take 20mL of lysis buffer (1:1 mix of 0.05N NaOH and 0.025 per cent SDS) in a microfuge tube.
2. Add 10mL of fresh Hemolymph and mix the contents in a vortex mixer.

3. Puncture of lid of the microfuge tube using a needle and then incubate in a waterbath at 100°C for 5 minutes.
4. After 5 min immediately place the tube on ice until it becomes cold.
5. Centrifuge the tube briefly and use 5mL of the supernatant for PCR analysis.

Advantages of PCR

1. PRC is relatively a very simple technique to perform.
2. The error probability, time consumption and cost are relatively less.
3. PCR can work efficiently even with poor quality templates. Formalin fixed specimens, paraffin embedded specimen and even crude cell lysates are found to provide excellent template for PCR.
4. The PCR technique is highly sensitive in the sense that it can detect and amplify from even few nanograms of the template. Its specified template will be amplified.
5. The PCR technique is very rapid and can be used to analyze larges number of samples within a short duration.
6. For sensitivity and specificity PCR is the ideal molecular based diagnostic currently available

Diagnosis of Diseases in Ornamental Fishes by PCR

PCR can be followed as a diagnostic tool in diagnosing various diseases affecting ornamental fishes. The following is the protocol to be followed for the diagnosis of bacterial, viral and parasitic disease affecting ornamental fishes.

Materials Needed

1. Fish samples in which the diagnosis has to be carried out -Live samples or samples fixed in ethyl alcohol can be used for PCR. Do not use dead or decayed samples.
2. DNA extraction chemicals or a commercial kit: Many standard methods are available for the extraction of DNA from the samples. It is easier to use commercial kits as they are simple
3. PCR master mix: A commercial PCR master mix with all the components (dNTPs, Buffer and Taq polymerase enzyme) in an optimized concentration would be the best choice.
4. PCR primers: Self-designed or published primers can be used for the diagnosis of various bacterial and viral diseases of ornamental fishes.

Equipment Needed for PCR

PCR thermal cycler, Micro centrifuge, Water bath, Weighing balance, Horizontal gel electrophoresis apparatus and UV transilluminator.

Steps Involved in the PCR Diagnosis of Diseases

1. Collection of samples
2. Extraction of Template DNA from samples
3. Setting up of PCR reaction
4. Amplification of template DNA using target specific in a PCR thermal cycler (Specific primers meant for each of the pathogen can be used)
5. Separation of the amplified DNA by gel electrophoresis
6. Observing the gel under UV illumination
7. Interpretation of results

PCR finds wide application not only in the diagnosis of diseases but also to study the epidemiology and prevalence of diseases, molecular mapping etc.

References

Felix, S. 2010. "*Marine and Aquaculture Biotechnology*". Agrobios (India), Behind Nasrani Cinema, Chopasani, Jodhpur. pp: 471.

Felix, S. 2008. "*Biosecured Aquaculture - Principle and Prototype*". Agrobios (India), Behind Nasrani Cinema, Chopasani, Jodhpur. pp:139.

Felix, S. 2008. "*Advances in Shrimp Aquaculture Management*". Daya Publishing House, Darya Ganj, New Delhi. pp: 192.

Chapter 19

Vaccine Development Strategies for Microbial Diseases

B. Samuel Masilamoni Ronald

Vaccine Research Centre-Bacterial Vaccine,
Centre for Animal Health Studies,
Tamil Nadu Veterinary and Animal Sciences University,
Madhavaram, Chennai – 600 051, T.N.

Aquaculture represents a real success story for agriculture and the farming of animals. The FAO predicts that by 2030 aquaculture will supply approximately one-third of the total world fishery. Indeed, aquaculture has many challenges ahead to meet this demand in a sustainable way. Though fish are not a significant source of human pathogens, farmed fish are susceptible to diseases. Pathogens include the usual suspects: parasites, virus and bacteria. To deal with disease, farmers and veterinarians require fish health management tools that are effective, environmentally responsible and acceptable to the public. The future success of aquaculture depends on developing and adopting technologies which allow efficient and environmentally sustainable production.

Current Status

Global aquaculture has grown dramatically over the past 50 years to around 52.5 million tonnes (68.3 million including aquatic plants) in 2008 worth US$98.5 billion (US$106 billion including aquatic plants) and accounting for around 50 per cent of the world's fish feed supply. Asia dominates this production, accounting for 89 per cent by volume and 79 per cent by value, with China by far the largest producer (32.7 million tonnes in 2008).

The average global annual growth rate of 7.6 percent (from 2000 to 2006) and the average annual growth rates of aquaculture producing countries in Asia suggest that all forecasted targets set for 2010 and 2020 by the forecast models are likely to be met.

Indian aquaculture has demonstrated a six fold growth in production over the last two decades, with freshwater aquaculture contributing the major share. The carp culture in freshwater and shrimp culture in brackish water forms the major activity in Indian aquaculture systems. In contrast, the development of brackish water aquaculture has been confined to a very few species, although India offers immense potential for the development of mariculture. The Department of Biotechnology demonstrated a semi-intensive culture and developed a technology to achieve the production level of 10 t/ha per annum in two crops in shrimp and more than 18-20 t/ha in fish production per annum through polyculture with introduction of high yielding hybrids, quality feed and breed and better health management. This has been a paradigm shift in fish and shrimp production in the country.

In aquatic health management, technology is being developed for screening healthy population of brooders based on molecular markers, maturation in captivity, completion of maturation in indoor facility wherever feasible, breeding and spawning in controlled systems. Novel concept on raceway technology using aerobic microbial flocculent technology is being propagated in designing, construction and operation of raceways to raise crops of various kinds including shrimp, scampi and freshwater ornamental fishes for future and to ensure the bio-security of used water for practicing sustainable aquaculture.

The diseases caused by infectious agents particularly viral and bacterial pathogens pose a major threat to a thriving aquaculture industry, and result in severe economic losses world-wide. The loss of shrimp production due to white spot syndrome virus (WSSV) and yellow head virus (YHV) has been estimated about US$1 billion per year in Asian countries (Adams *et al.*, 2008). In India, the loss of shrimp due to WSSV has been estimated about US$ 1.50 billion per year (CIBA, 2008). Application of various technologies especially biotechnological tools has made an impact in reducing disease risk and advance technologies are contributing to the future enhancement of aquaculture production. Improved nutrition, use of probiotics, improved disease resistance strain, water quality, seed and feed, use of immuno-stimulants, rapid detection of pathogens and the use of affordable vaccines have all assisted in health control measures in aquaculture. The rapid diagnostic method is helpful in detection of pathogens in different situations such as in clinically infected animals, in sub-clinically infected animals or in the environment. There is an urgent need to develop a simple, rapid, cost effective, sensitive detection tool for diagnosis of various pathogens through innovation methods and addressing fish health management issues and bio-safety aspects in aquaculture.

The Department of Biotechnology is addressing various aquatic health issues through establishment of diagnostic laboratories to screen the shrimp brooders and seeds for viral pathogens using molecular tools such as PCR, RT-PCR and immunodiagnostics through the projects supported at various aquaculture institutes in the country. Support is also provided to the programmes like production SPF brooders and seeds using molecular tools including RNAi technology and developing brood stock banks to supply of high quality brooders to hatchery operators. Fish cell lines and shrimp primary cell culture, the outcome of DBT-funded projects, are being used for propagation of viruses for production whole virus vaccines. There is a need of National Referral Facility to take care of aquatic animal health by providing diagnostic facility and control and preventive measures for all bacterial and viral diseases encountered in Indian aquaculture system.

In all forms of intensive culture, where single or multiple species are reared at high densities, infectious disease agents are easily transmitted between individuals. In general most of the bacterial diseases that affect ornamental fish are caused by opportunistic bacteria. Which means that most bacterial infections result because of changes in the bacteria/fish relationship. This is a very delicate relationship, but generally when an opportunistic bacterium tries to establish an infection the fish host is able to resist by a variety of defence mechanisms. A substantial increase in the numbers of opportunistic bacteria such as we would find in systems with other infected fish and/or high levels of decomposing organic matter. This is typically the situation when the disease starts to spread and several fish are affected. Under these conditions, especially if fish are still stressed, it seems that some opportunistic bacteria can become primary pathogens through weight of numbers.

It is because of these complications, together with tremendous reproductive potential of most bacteria, that a seemingly minor problem can rapidly snowball out of control leading to high numbers of infected fish and possibly many losses.

Fish Immunology

All vertebrates have mechanisms for controlling pathogens – those organisms which are capable of causing disease. However fish are the most primitive organisms to have an adaptive immune system which is comparatively simple and undifferentiated compared with mammals. The immune system of fish has evolved with both non-specific (innate immunity) and acquired immune functions (humoral and cell mediated immunity) to eliminate invading foreign living and non-living agents.

Innate mechanisms require no previous exposure to the particular agent – this includes physical barriers such as skin and mucus layers, specialized cells such as macrophages and natural killer cells and particular soluble molecules such as complement and interferon. The first line of defence which fish have against foreign agents are mucus and skin, contain immuno-reactive molecules. Recently specific antibody to parasites and bacteria were demonstrated in mucus. Non-specific humoral molecules in fish include lectins (carbohydrate recognition), lytic enzymes, transferrin (iron binding protein) and components of the complement system. Non-specific cells of the fish immune system include monocytes or tissue macrophages, granulocytes (neutrophils) and cytotoxic cells. Macrophages function in phagocytosis and destruction of invading foreign agents and bacteria.

Acquired immunity in fish includes both humoral and cell mediated responses. Fish can display typical vertebrate adaptive immune responses characterized by immunoglobulins, T-cell receptors, cytokines, and major histocompatibility complex molecules (HSC). The cell-mediated response in fish is similar to that in mammals and relies on the presence of accessory cells (macrophages) to present antigen to T-cells. The correct presentation of antigen results in a cascade of events that includes cytokine production that regulates or enhances the cellular response.

The anterior portion of teleost fish kidney is most likely the source of HSCs that will later give rise to the B and T-cell lineages. In teleleost fish, progenitor T-cells migrate from the kidney to the thymus for Tcell education (distinguishing self from non-self) and maturation (functional). The B-lymphocytes originate and mature within the kidney, therefore the anterior region of the fish kidney is considered to be the evolutionary equivalent of the bone marrow. B-cells of fish produce antibody when stimulated.

One major difference is that fish in general are poikilothermic – in other words they adapt their body temperature to that of their surroundings – *i.e.* the water temperature and their metabolic rates and development of immune response are therefore directly dependent on the temperature of their aquatic environment. Therefore, as with mammals, when a fish encounters an infectious agent, its response will depend on whether it has experienced this infectious agent previously. Aspects of both the innate response and the adaptive immune system will come into play depending on the type of infectious agent, route of infection and history of previous contact with this infectious agent. Other factors such as stress and nutrition status will also have a part to play in the quality of the response.

History of Fish Vaccinology

The first vaccines developed in the 1970's and 1980's were water based formalin-killed vaccines for dip immersion, using strong antigens such as *Yersinia ruckeri* and *Vibrio anguillarum*, which were applied by intra-peritoneal injection to larger fish. To improve the efficacy of such vaccines, especially when using weaker antigens such as *Photobacterium damsela* subsp. *piscicida*, ultrason were tested with immersion vaccination (Navot *et al.*, 2005) and mineral or non mineral oil adjuvants were applied with injectable vaccines. However, side effects rapidly became an issue, which was partly solved by using non-mineral oil adjuvant. The development of effective vaccines for oral delivery represents the last step in the development of commercial vaccines. The licensing of an oral enteric redmouth (ERM) vaccine was the first full license ever to be granted for an oral fish vaccine, but it not only has an effect on ERM vaccination strategies but establishes the credibility of the method as well. Proving the technology will have a profound effect on the vaccination strategies for other vaccines. Other antigens, such as *Vibrio anguillarum* serotypes I and II in Greece or *Lactococcus garviae* in Japan are already available for oral delivery in marine species. During the last few years, the use of vaccines for disease prevention in aquaculture has expanded both with regard to the number of fish species and number of microbial diseases. Vaccines used in the commercial aquaculture of species like Atlantic salmon (*Salmo salar*), rainbow trout (*Oncorhynchus mykiss*), sea bass (*Dicentrarchus labrax*), sea bream (*Sparus aurata*), barramundi (*Lates calcarifer*), tilapia (*Tilapia spp*), turbot (*Scophthalmus maximus L.*), yellowtail (*Seriola quinqueradiata*), purplish and gold-striped amberjack

(*Seriola dumereli*), striped jack (*Pseudocaranx dentex*) and channel catfish (*Ictalurus punctatus*). The range of bacterial infections for which vaccines are commercially available now comprises classical vibriosis (*Listonella anguillarum, Vibrio ordalii*), furunculosis (*Aeromonas salmonicida subsp. salmonicida*), cold-water vibriosis (*Vibrio salmonicida*), yersiniosis (*Yersinia ruckeri*), pasteurellosis (*Photobacterium damselae supsp. piscicida*), edwardsiellosis (*Edwardsiella ictaluri*), winter ulcer (*Moritella viscosa*), and streptococcosis/lactococcosis (*Streptococcus iniae, Lactococcus garviae*). Furthermore, experimental vaccines are used against diseases such as infection with *Vibrio harveyi* and *Photobacterium damsela subsp. damsela* in barramundi, piscirickettsiosis and bacterial kidney disease in salmonids, as well as infection with *Flexibacter maritimus* (now: *Tenacibaculum maritimum*) in turbot.

Environmental Perspective

Environmental bodies and organizations are putting more and more emphasis on pollution resulting from aquaculture activities. It is likely that restrictive regulations against indiscriminate use of antibiotics will be enforced. Vaccination therefore becomes an alternative, requested by an increasing number of customers and already included in the list of criteria of different quality schemes and labels: (i) hyper market quality chart in Spain; and (ii) red label and organic label in France.

Types of Vaccines

Vaccine types used in farmed fish can be broadly divided as follows:-

1. Inactivated bacterial vaccine
2. Inactivated viral vaccine
3. Subunit vaccine (derived from recombinant technology)
4. DNA vaccine

i. Inactivated Bacterial Vaccines

This vaccine is prepared by producing bulk cultures and inactivating with formalin which kill all of the bacteria while preserving their antigenicity. This inactivated culture is further adjuvanated with adjuvants for improved and prolonged antigen response.

ii. Inactivated Viral Vaccines

This vaccine is prepared by raising bulk quantities of virus in tissue culture and killing with inactivating agents that will not alter the antigenicity of the virus. The inactivated culture is mixed with adjuvant for improved and prolonged antigen presentation for better immune response.

iii. Subunit Vaccines

Subunit vaccines are produced by implanting the part of the DNA which encodes for the production of the specific antigens to trigger an adequate immune response into another type of organism. The recipient organism is chosen for characteristics such as the ability to be easily produced on a large commercial scale and also to secrete the donor protein antigens in a format which can be readily commercially harvested.

iv. DNA Vaccines

The principle of DNA vaccination is to make the host animal receiving the vaccine responsible for the production of the antigen and then the immune response to the antigen. It is also known as Genetic immunisation. Basically DNA vaccination or nucleic acid immunisation entails the delivery of DNA (or RNA) encoding a vaccine antigen to the recipient. The DNA is taken up by host cells and transcribed to mRNA, from which the vaccine proteins are then translated. The expressed proteins are recognised as foreign by the host immune system and elicit an immune response, which may have both cell-mediated and humoral components. DNA vaccines offer a number of advantages over conventional vaccines, including ease of production, stability and cost. They also allow the production of vaccines against organisms which are difficult or dangerous to culture in the laboratory.

Development of Vaccination Strategies

The early fish vaccines were very basic formalin inactivated bacterial cultures, which were administered initially by immersion and subsequently by injection. The immersion technique has endured for some bacterial antigens since it is still the most effective method of ensuring short-term immunity to small fish which for a number of reasons including size, individual value and reaction to the general stress of handling make ideal candidates for this approach. Although the immersion technique certainly produced some levels of immunity to certain bacteria, further research was directed at the injectable vaccines. Although a degree of protection can be achieved by simply injecting the killed bacterial suspension i.p., improved levels and duration of protection were dependent on the development of suitable adjuvants. However the early oil-based adjuvants had the undesirable effect of producing lesions within the peritoneal cavity post-vaccination.

Oral vaccine technology has been seen over the years as the ideal method of administration of vaccines to fish. The major limiting factors in oral vaccination are the presentation of the vaccine in a suitable form, ensuring even distribution of the dosage form throughout the population of fish to be vaccinated, duration of immunity and presentation of the appropriate range of antigens.

In summary, a great deal of development is required to produce a safe and efficacious vaccine. The detail and course of this varies greatly depending on the type of disease, and the nature of the vaccine and target stock.

Methods of Vaccine Administration

1. Oral
2. Injection
3. Immersion/Spray

1. Oral

The method of oral administration can vary according to the vaccine. The three methods are

1. Top-dressing the finished feed with the vaccine powder using an adhesive agent such as edible oil or even gelatine,
2. Spray-dressing the finished feed if the vaccine is in liquid form,
3. Incorporating the vaccine into the feed during the feed manufacturing process.

The biomass of the fish to be vaccinated should be estimated and the vaccine mixed with the feed according to the manufacturer's instructions. With liquid vaccines, bring the vaccine to room temperature (20°C) for 1 hour before use to allow the vaccine to become more liquid. If any separation occurs, shake the bottle vigorously until the separated layers are completely dispersed. Turn the required weight of feed pellets in a mixer, *e.g.* a concrete mixer, and slowly pour or spray the vaccine directly onto the pellets. If a sprayer is used, it should be set to deliver a coarse spray without risk of aerosol particle generation and the spray container must be completely emptied during the mixing operation. Mix the pellets for at least 2 minutes after all the vaccine has been added. Keep the prepared feed for 1 hour before feeding, to allow the vaccine to impregnate the pellets completely.

2. Injection

If the fish is sufficiently large to be individually handled (>50g average weight) the vaccine may be safely administered undiluted by injecting them intraperitoneally using injection guns. Groups of fish are cut off from the rest in a cage and enclosed in a tarpaulin as for immersion, but because the fish are larger, a smaller number of them is enclosed. Extra care is also required when supplying the anaesthetic since the larger the fish the greater the risk of self-injury due to stress reactions. Subsequent to sedation in the tarpaulin, small groups of fish are netted out and placed in a container with a higher dose of anaesthetic dilution until completely immobile.

The anaesthetised fish are then taken to a "vaccination table" where they are handled individually. The vaccination table consists of one or more troughs filled with water where the immobile fish are presented to the operators floating belly up. A certain dose of vaccine (usually 0.1ml to 0.2ml) is injected in the abdominal area of each fish held with the ventral side up and the head away from the operator´s body. The needle is inserted into the peritoneal cavity at a 45° angle to a depth of approximately 0.5 cm. Automatic injection guns are used for this purpose.

Subsequent to injection the fish are released into their holding unit where they recover from anaesthesia in a few minutes. Usually, the vaccination table is constructed in a way that allows simultaneous easy grading of the injected fish into size groups. The fish are released according to size class into distinct channels. Water is pumped from the sea into the channels and flushes the fish along tubes leading them to different cages. The table, pumps, tubing and all other implements are often installed on a floating platform.

3. Immersion Vaccination

Large groups of fish are cut off from the rest in a cage and enclosed in a tarpaulin where a slight dose of diluted anaesthetic is added to sedate them. Air or oxygen is continuously pumped in to avoid anoxia.

The proper vaccine quantity is calculated according to the estimated biomass of the fish to be vaccinated (usually for every 100kg of fish one litre of vaccine is required). The vaccine is diluted with sea-water in a suitable receptacle (the common dilution rate is 1:10, that is, 9lt of water added for each 1lt of vaccine). Oxygen may advantageously be trickled through the vaccine dilution to reduce stress. The sedated fish are netted out of the tarpaulin in lots of approximately 0.5kg, avoiding overcrowding or crushing. Holding water is drained from the fish, which are then placed in the vaccine dilution where they are allowed to swim for a certain minimum time, usually for 30 seconds.

Common practice is to place the netted fish from the tarpaulin in a perforated plastic bowl within the dilution container. When immersion time is up, the fish are withdrawn from the vaccine dilution and released in their holding facility. Groups of caged fish are cut-off in a tarpaulin where oxygenation is provided and a light dilution of anaesthetic is added for sedation. The fish are netted out in small groups, drained for a few seconds from sea water and immersed in the dilution of vaccine. Usually a perforated plastic container is used to hold the fish in the vaccinal dilution.

The sedated fish remain in the perforated bowls for at least 30sec. Then the bowl is up-lifted from the vaccine dilution and is left to drain for a few seconds prior to transferring the vaccinated fish to their cage or raceway. After drainage, the vaccinated fish are carefully released into the receptor raceway or cage, situated next to where the immersion takes place.

Table 19.1: Advantages and Disadvantages of Vaccination Methods

Vaccination Method	*Advantages*	*Disadvantages*
Immersion vaccination (1) Dip immersion (2) Long term bath (3) Spraying	☆ Suitable for large quantities of small fish (<5 gr) ☆ Cost effective for small fish (1 and 3) ☆ Protective immunity for 3-5 months depending on the antigens	Difficult to apply on ongrowing units Expensive for large fish Costly (2)
Injection vaccination I.P.or I.M.	☆ Good protective immunity, lasting up to 1 year ☆ Suitable for large fish (broodstocks) ☆ Possibility of automatization (tables) ☆ Good for weak antigens, for vaccination at low temperature	☆ Stressful method ☆ Labor and time consuming ☆ Use of anesthetic required ☆ Side lesions
Oral vaccination	☆ No stress, not time consuming ☆ Easy to apply on all production facilities ☆ For all sizes o ffish (>10 g) and large batches	☆ Monovalent vaccines only ☆ Protection slightly shorter than injection (8 months) ☆ Needs to be planned and requires food feeding practice

These three methods of vaccination are further dependent on different production factors:

1. The economic factor represents the bottom line. Vaccination will be applied if the cost of the disease is expected to be higher than the cost of the vaccination.
2. The size of the fish will largely define if immersion vaccination is suitable or not compared to injection or oral delivery.
3. The production structure and production management of the site considered will be a factor mostly for booster vaccination and the choice between injection or oral delivery of the vaccine.
4. The period of the year and the species to be immunized will also influence the choice of the booster vaccination method.

Factors Influencing the Vaccination Strategy

To be successful, a vaccination strategy has to be adapted to the site where it will be implemented, depending on different factors:

1. The Epidemiology of the Diseases on the Site

A proper identification of the pathogen(s) threatening the production on site is an essential primary step.

2. The Farm Production Strategy and Management

A minimum period, depending on the water temperature, is required for the development of the specific immune response of the fish.

3. The Species

Each fish species is sensitive to a specific range of pathogens and this sensitivity can change during the different stages of their life cycle. These parameters will guide the choice of the vaccine, while the method of vaccination to apply will depend on the stress sensitivity of the species to vaccinate.

4. The Hygiene on Site and the Health Status of the Fish

The health status of the fish is fundamental for successful vaccination. Gill parasitic infestationor bacterial gill disease will not allow a good uptake of antigen through this organ during immersion vaccination. For the same reason, anesthesia of the fish slows down the blood flow through the gill filaments and represents a contra-indication for vaccination of the fish by immersion.

Conclusion

Commercial vaccines, which have proved their efficacy under laboratory conditions and in field trials, are available for the main bacterial diseases threatening aquaculture production facilities. Therefore, the claims of unsuccessful vaccination using a commercial vaccine should be investigated with correct diagnosis of the resulting mortality. This approach shows in most cases that the vaccine itself is not at fault, since the vaccination strategy applied was not properly adapted to the farming conditions on site, to the disease considered and its epidemiology or the species needing protection. Vaccination programs represent only one part of health management and should be applied in parallel with other prophylactic methods, such as sanitary rules to optimize production results.

Index

Coloured Plates

Figure 1.1: Some of Marine Sponges and Seaweeds Collected from Indian Coast and Tested for Bioactivity and Aquaculture Drugs. (Page 6)

Figure 4.5: Submerged Culture Method (Page 46)

Figure 4.6: Surface Culture Method (Page 46)

Figure 6.10: Seeds being Acclimatized (Page 73)

Figure 6.12: Fermentation Chamber (Page 77)

Figure 6.13: Indoor Algal Culture Facility (Page 78)

Figure 10.1: Infusoria Culture. (Page 120)

Figure 10.2: Hatching of Artemia Cysts. (Page 121)

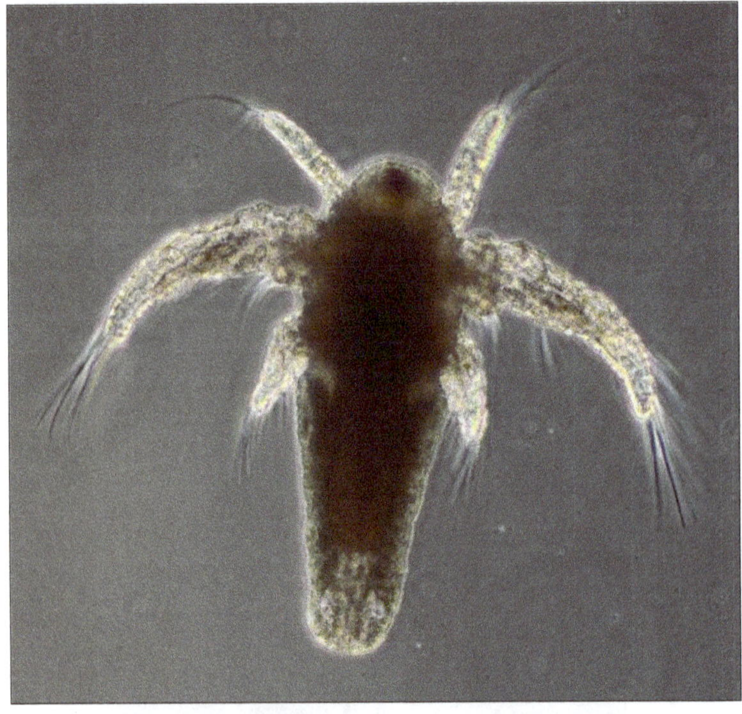

Figure 10.3: Hatched Out Artemia Nauplii. (Page 122)

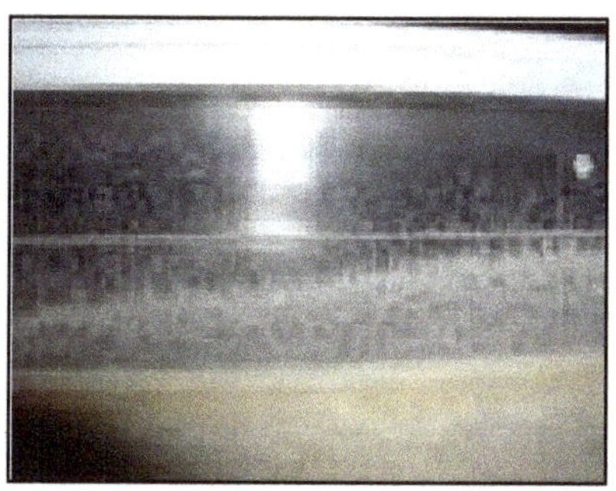

Figure 10.4: Microworms. (Page 124)

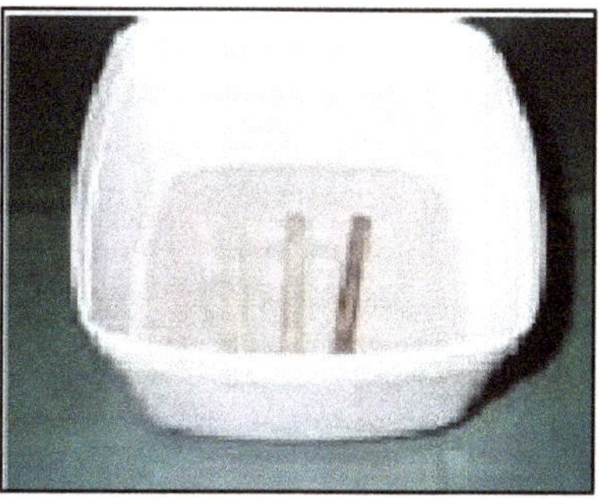

Figure 10.5: Culture of Microworms. (Page 125)

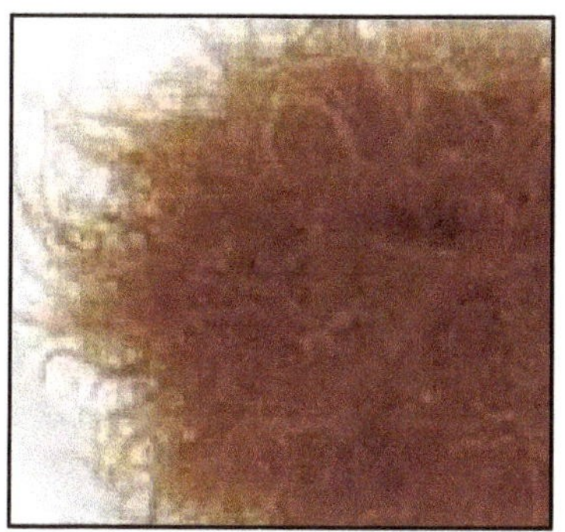

Figure 10.6: Tubifex Worms. (Page 127)

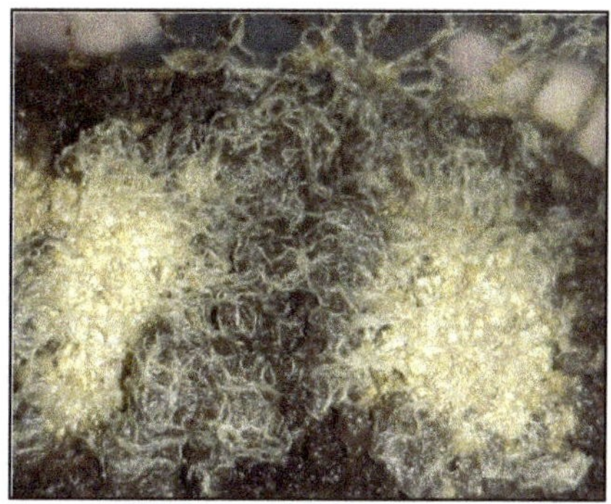

Figure 10.8: White Worms. (Page 129)

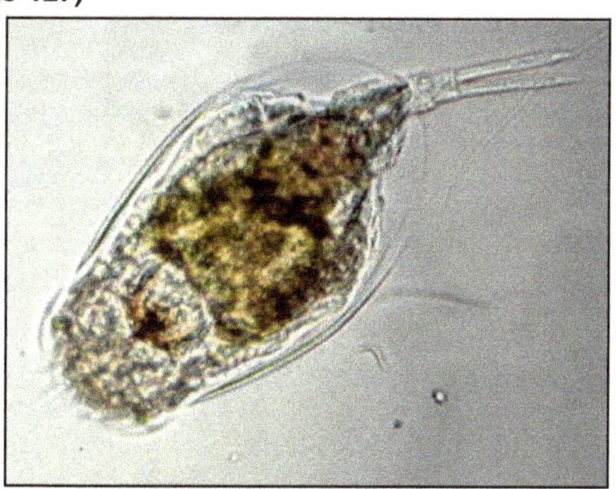

Figure 10.9: Rotifer (*Brachionus* sp.). (Page 132)

Figure 15.1: Goat TLR 1 Protein Domain Predicted by SMART Analysis (Page 198)

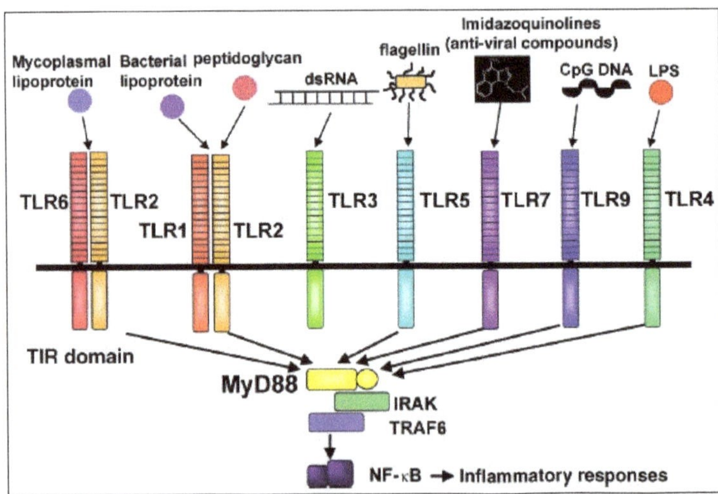

Figure 15.2: TLRs Recognize Molecular Pattern Associated with Bacterial Pathogens. Triacylated Lipoprotein for TLR1; Peptidoglycan for TLR2; double-stranded RNA for TLR3; lipopolysaccharide (LPS) for TLR4; flagellin for TLR5; diacylated lipoprotein for TLR6; imidazoquinoline and its derivative R-848, for TLR7; and bacterial unmethylated CpG DNA for TLR9. MyD88 associates with the TIR domain of TLRs and transduces signals to induce immune responses. (Page 199)

Figure 16.2: Starter Culture. (Page 210)

Figure 16.3: Flask Culture (Page 211)

Figure 16.4: Algal Culturing at Room Temperature. (Page 211)

Figure 16.5: Chlorination of Seawater. (Page 212)

Figure 16.6: Outdoor Algal Culture (20 lit.) (Page 212)

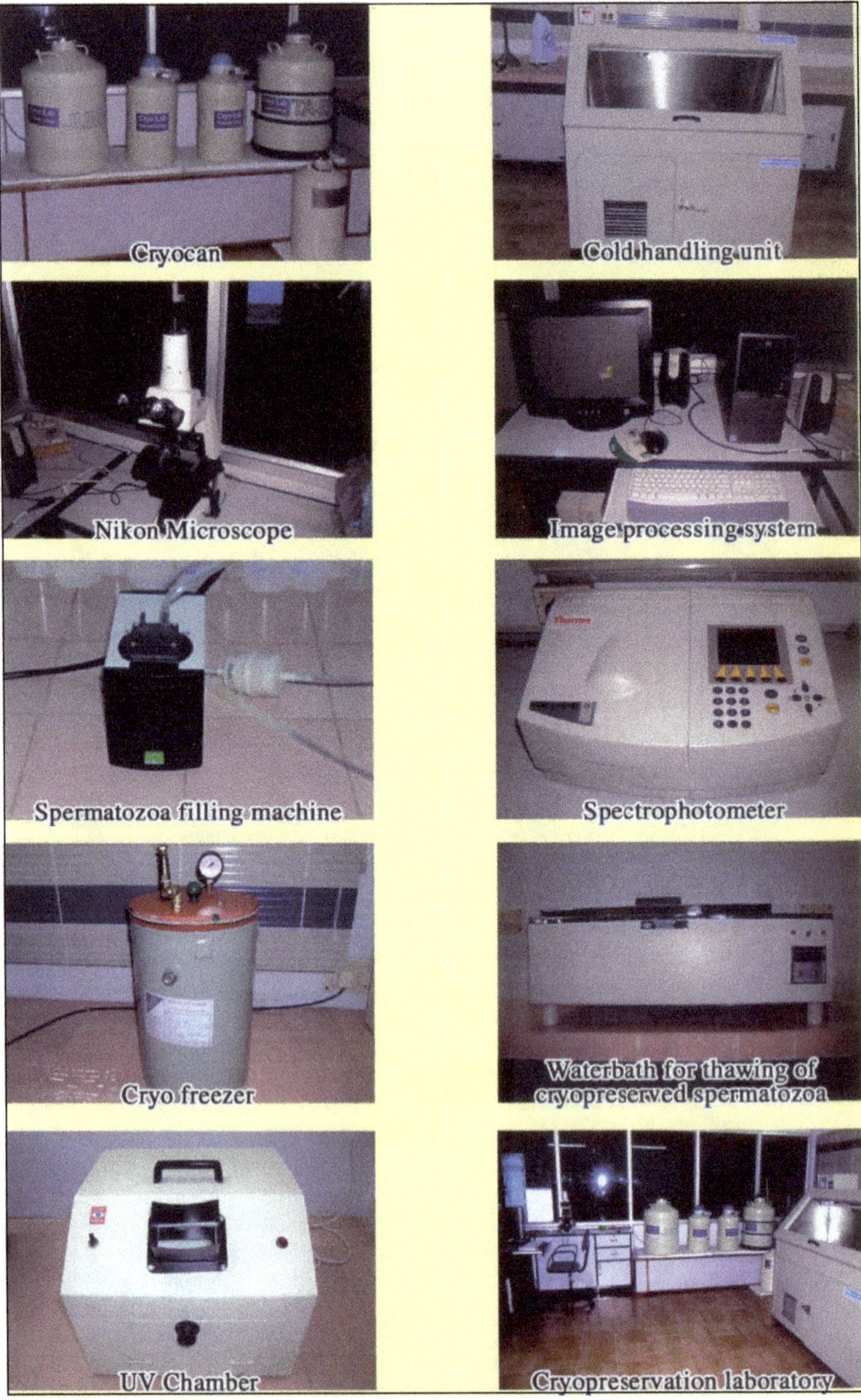

Figure 17.2: Lab Facility for Cryopreservation of Fish Spermatoza (Page 227)

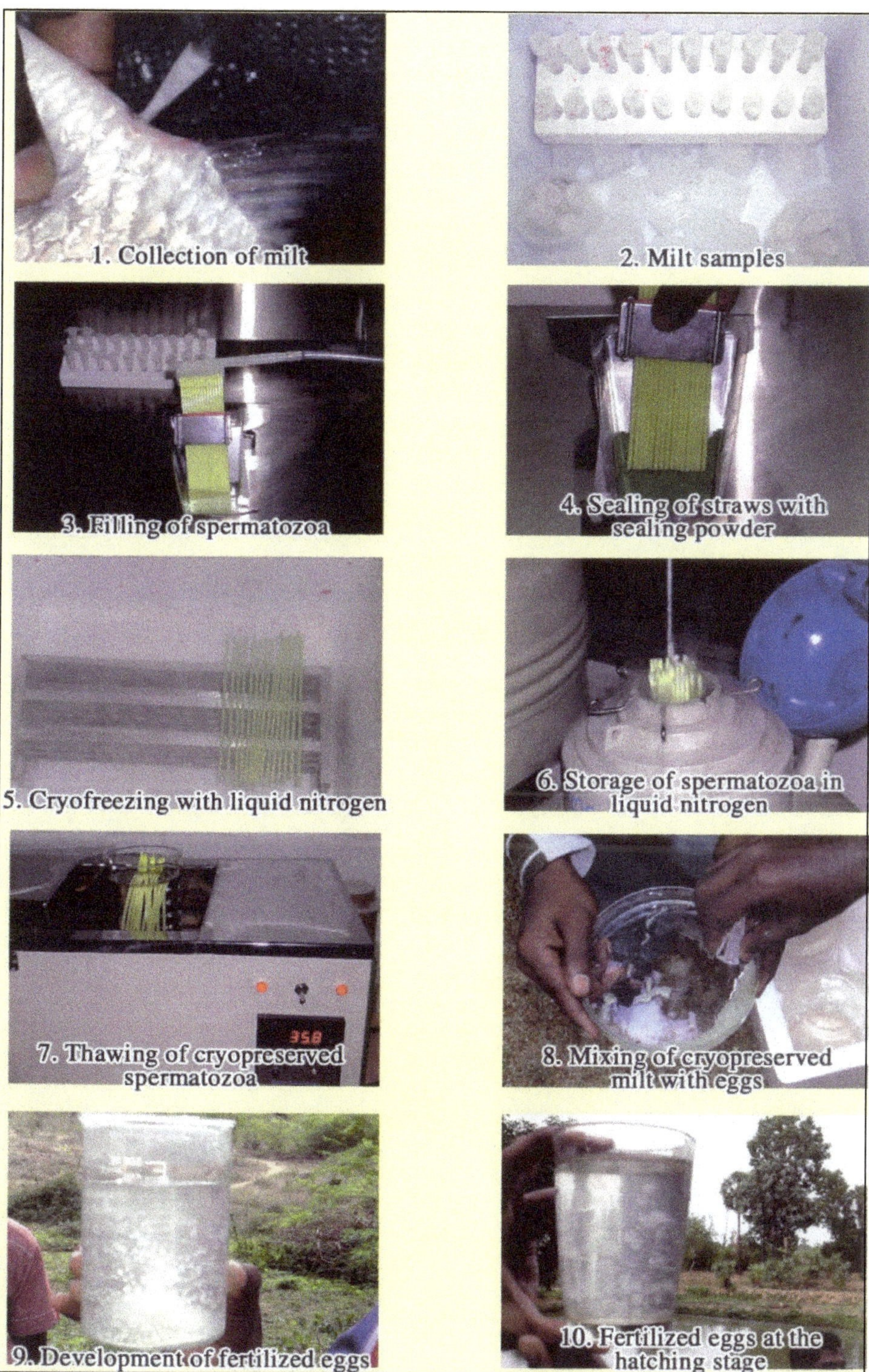

Figure 17.3: Cryopreservation Protocol (Page 228)

www.ingramcontent.com/pod-product-compliance
Ingram Content Group UK Ltd.
Pitfield, Milton Keynes, MK11 3LW, UK
UKHW021010290726
14059UKWH00001BA/53